Newton's Metaphysics of Substance

Newton's Metaphysics of Substance

God, Bodies, Minds

PATRICK J. CONNOLLY

OXFORD
UNIVERSITY PRESS

OXFORD
UNIVERSITY PRESS

Great Clarendon Street, Oxford, OX2 6DP,
United Kingdom

Oxford University Press is a department of the University of Oxford.
It furthers the University's objective of excellence in research, scholarship,
and education by publishing worldwide. Oxford is a registered trade mark of
Oxford University Press in the UK and in certain other countries.

Published in the United States of America by Oxford University Press
198 Madison Avenue, New York, NY 10016, United States of America.

British Library Cataloguing in Publication Data
Data available

Library of Congress Control Number: 2025942276

ISBN 9780192867117

DOI: 10.1093/9780191959240.001.0001

The manufacturer's authorised representative in the EU for product safety is
Oxford University Press España S.A. of Parque Empresarial San Fernando de Henares,
Avenida de Castilla, 2 – 28830 Madrid (www.oup.es/en or product.safety@oup.com).
OUP España S.A. also acts as importer into Spain of products made by the manufacturer.

For Elanor

Contents

Acknowledgments

In the course of writing this book, I have benefitted enormously from the time, energy, and expertise of many individuals as well as the resources and support of many institutions. I wish to thank them here.

The deep origins of this project lie in a graduate seminar I took as an MA student in the spring of 2009. The seminar was on Leibniz and Newton and was co-taught by Alan Nelson and Andrew Janiak. Their ability to make Newton's philosophical thought accessible, exciting, appealing, and rich is what has led, ultimately, to this book. I am grateful to them for that. But I am also grateful to them for much else. Throughout my graduate studies, the earlier stages of my career, and into the present, they have both been unflagging sources of knowledge, support, advice, and care. I have learned so much from both of them. This book is deeply indebted to their efforts.

Very special thanks also go to Zvi Biener, Lisa Downing, Geoff Gorham, and Ed Slowik, all of whom read an earlier version of the manuscript in its entirety and traveled to Baltimore to discuss it with me over two days in September of 2024. Katherine Brading and Andrew Janiak, while unable to join in person, also read the manuscript and offered valuable feedback at that time. I am grateful to all six of these scholars for their generous attention, deep insights, careful questions, and principled objections. They saved me from a number of errors, enriched my understanding of Newton and his context, improved my arguments, and greatly aided in the revision of the text. Similarly, I would like to thank an anonymous referee for OUP who read the manuscript and shared their expertise by offering a series of insightful comments and suggestions for improvement.

I have spent much of the past several years discussing aspects of this book with anyone and everyone willing to listen. In addition to those already mentioned and those mentioned below, and with apologies to anyone I may have inadvertently omitted, my heartfelt thanks go to Peter Achinstein, Jean-Pascal Anfray, Sebastian Bender, Jip van Besouw, Ruth Boeker, Colin Chamberlain, Graham Clay, Hao Dong, Mark Fiocco, Jim Force, Dan Garber, Hüseyin Güngör, Scott Harkema, John Henry, Karolina Hübner, Michael Jacovides, Julia Jorati, Vili Lähteenmäki, Jacob Lettie, Alberto Luis López, Yitzhak Melamed, David Miller, Steve Nadler, Jen Nguyen, Eve-Lyne Perron, Jasper Reid, Sam Rickless, Robert Rynasciewicz, Anat Schechtman, Brian Simboli, Justin Steinberg, Katie Tabb, Gabe Watts, Shelley Weinberg, John Whipple, and Isaac Wilkins.

Earlier versions of chapters were presented as talks at the Theoretical Philosophy Colloquium at Humboldt University of Berlin, the New York City Seminar in Early

Modern Philosophy hosted by Technion-Israel Institute of Technology (both of which, due to the COVID pandemic, I had to attend on Zoom), the 2021 Northwest Philosophy Conference (where I received helpful comments from Graham Hubbs), the 2022 Eastern APA in Baltimore (where I received helpful comments from Areins Pelayo), the Philosophy Club at Loyola University Maryland, and the Chicago Modern Philosophy Roundtable hosted by the University of Illinois, Chicago. My thanks to all of those who made these presentations possible and to the audience on each occasion for their time, questions, and feedback.

Material central to this book was also presented in a variety of seminar formats. I would like to thank the students in my undergraduate seminar on Newton and Descartes in the spring of 2018. Jeremy Ahn, Shane Del Bianco, Brandon Diaz, Shazeb Fahim, Christie Furman, Yan Jin, Emily Krulik, Zach Sokol, and Chris Szafranski all showed enormous good will in delving into Newton with me and I profited a great deal from our class discussions. I am also grateful to the members of the Lehigh Philosophy Faculty Seminar in the fall of 2018: Gordon Bearn, Mark Bickhard, Ricki Bliss, Filippo Casati, Robin Dillon, John Gillroy, Chad Kautzer, Greg Reihman, Brian Simboli, and Roslyn Weiss. They were enormously patient as I began to find my way through this material and helped with the initial formulation of key ideas in the book. Finally, I am grateful to the graduate students who enrolled in my seminar on Newton in the fall of 2023: Elliot Mertz, Connor Morrison, Emily Schofield, and Madeleine Schmitz. Our discussions deeply enriched my understanding of a number of points central to the book.

Material included in Chapter 3, Section 3.6, previously appeared in secs. 2 and 3 of my 2015 article "Space Before God? A Problem in Newton's Metaphysics" *Philosophy* 90(1): 83–106. The article is © Cambridge University Press and the content is reprinted with permission. My thanks to the press for permission to use the material here.

During the years I worked on this book, I was affiliated with five different institutions. I consider myself extremely fortunate to have been a member of the rich intellectual community at each of them. It is a pleasure to recall and acknowledge the encouragement and support I received from the faculty, staff, and librarians at Lehigh University, Humboldt University of Berlin, the University of Wisconsin-Madison, Johns Hopkins University, and the National Humanities Center. My time at Humboldt University of Berlin was funded by a generous fellowship from the Alexander von Humboldt Foundation. Very simply put, this book would not have been written without the time and support provided by the Humboldt fellowship. Further, the fellowship would not have been possible without the support of Dominik Perler, my fellowship sponsor and academic host in Berlin. I am grateful to Dominik for the enthusiasm he showed for my project and his positive contributions to it. More generally, I am grateful to him for the knowledge, care, skill, and wonderful energy he brings to every philosophical interaction. He is a model philosopher, an inspiration, and a *sine qua non* cause of this book. I am deeply grateful.

I would also like to thank my family. I am incredibly lucky to have had loving and supportive parents, Jim and Coleen, who have always believed in me and have always been eager to hear about my academic pursuits. My sisters—Bridget, Rose, and Molly—have been equally wonderful, especially when times have been hardest. Finally, my biggest thanks go to Elanor Taylor. Philosophically and otherwise, there is no one to whom I am more indebted. This book is dedicated to her.

1
Introduction

1.1. Introduction

This book aims to reconstruct and explore Newton's views on a number of foundational questions relating to the metaphysics of substance. Of course, Newton never wrote a strictly philosophical treatise explaining his views on these issues. My aim, however, is to show that Newton was in fact well-positioned to have done so and that we are all the poorer for his not having done so. This is because, while he stopped short of presenting his views in a developed and systematic way, Newton devoted considerable thought to these issues. And the product of his careful contemplation was an understanding of God, created substances, and their relations with one another that was innovative, perceptive, and—by the standards of the period—compelling.

This introduction will first begin with a brief overview of some moments in Newton's biography that are particularly relevant for the subject matter of the book. It will then ask about the various domains to which Newton applied himself and how we should see them as relating or not relating to one another. This will help to offer an understanding of where metaphysics might fit in his thought, specifically how we might think about its relationship to his natural philosophical theory and practice. I then articulate my understanding of metaphysics and the sense in which it may or may not have been an actor's category for Newton. Following that, I offer some brief remarks about Newton's broadly empiricist methodology before turning to issues relating to sources and chronology. I conclude with an outline of the subsequent chapters in the book and give a brief summary of their contents.

1.2. The Metaphysics of Substance and Newton's Intellectual Biography

It may be useful to begin with some overview of Newton's metaphysical thought in the context of his life and career. What, for instance, was Newton's "philosophical upbringing"? With which philosophers did he engage? What kinds of questions might he have been exposed to? At what stages in his life was he particularly oriented toward metaphysical issues? This section, while stopping short of providing full answers to any of these questions, attempts to provide a starting point from which they can be considered.

Newton's Metaphysics of Substance. Patrick J. Connolly, Oxford University Press. © Patrick J. Connolly 2025.
DOI: 10.1093/9780191959240.003.0001

Newton's intellectual career began during his childhood and adolescence in the Lincolnshire of the late 1640s and the 1650s.[1] From an early age, Newton evidenced deep curiosity about the natural world and tremendous facility with what we might call the mechanical arts, building models, toys, and small machines. He pored over works like John Bate's *Mysteries of Nature and Art* and John Wilkins' *Mathematicall Magick*. While it would be a mischaracterization to describe either of these as being oriented toward or offering treatments of metaphysics, both certainly engaged with or referred to metaphysical topics that Newton came to devote serious attention to in subsequent years. Bate, for instance, appeals to both philosophical authority and empirical evidence to deny the existence of vacua or empty spaces in nature.[2] In the course of laying down "some few Principles or Præcognita" that might enable readers "to know the reasons and causes" of things, Bate stresses the metaphysical principle that: "It is impossible for one and the selfe same body to possesse at one time two places" and traces through the implications of this claim for condensation and rarefaction.[3] Wilkins' text also contains a variety of substantive metaphysical claims. Like Bate, he affirms the impossibility of a vacuum and goes on to deny that two bodies can co-locate.[4] He writes that "a materiall substance is altogether incapable of so great a celerity, as is usually ascribed to the celestiall orbs" and that following on from a certain chymical procedure "there will bee at length produced divers small atomes, which being put into a glass well luted, and kept dry, will have perpetuall motion."[5] It is, of course, impossible to know whether the very young Newton lingered over these passages or spent significant time trying to puzzle through them. But, given his careful attention to related issues in later years, they may well mark the beginning of his fascination with metaphysical issues.

At the age of twelve, Newton began attending the King's School in Grantham. It is unlikely that his education there would have included very much, if anything at all, by way of training in what we would now call metaphysics. But it is worth noting certain links between this school and the prominent Cambridge metaphysician Henry More, whose work Newton would later come to know well and appreciate. More himself was a native of Grantham and had attended the school. While Newton was a student, a certain Dr. Joseph Clark, another graduate of the school, was one of the assistant masters. Clark had been one of More's pupils at Cambridge. As a student, Newton lodged in the house of Clark's brother. It also seems that More stayed with the Clarks during his visits to Grantham.[6] Of course,

[1] Again, this section offers only a very brief and selective overview of Newton's philosophical biography. Westfall 1980 remains a remarkable resource for understanding Newton's life and career. Further sources are mentioned below as relevant.

[2] Bate 1634, sig. A3.

[3] Bate 1634, pages 53–54. Gabbey 2011, page 424, draws attention to this discussion in connection with Newton.

[4] Wilkins 1648, page 250.

[5] Wilkins 1648, pages 141–142 and 228.

[6] On these points about Clark, see Reid 2012, page 2.

none of this entails that as a schoolboy Newton came across—let alone immersed himself in—metaphysical discussions influenced by More. But, as with Bate and Wilkins, it is possible he was exposed to something that piqued his curiosity or laid the foundations for his future pursuits.

One other feature of Newton's life in Lincolnshire deserves mention in connection with his systematic metaphysics. Newton's stepfather, Barnabas Smith, was the rector at North Witham. Upon his death in 1653, Newton seems to have inherited his library. As Westfall writes, "Newton's room at Woolsthorpe contained, on the shelves that Newton had built for them, two or three hundred books, mostly editions of the Fathers and theological treatises."[7] We know that Newton certainly made use of Smith's mostly blank theological commonplace book; it became his "Waste Book", filled with important mathematical and mechanical notes. Whether the precocious young Newton also found time to engage with the published works received from his stepfather is less clear. His most devoted attention to Patristics seems to have come later, in the 1680s and early 1690s.[8] But I think we ought not discount the possibility that he explored this collection in at least some depth. If he had, he would have found a wealth of rich metaphysical reflection on topics that fascinated him throughout his life.

Here is just one example. Newton owned the 1638 edition of Gregory of Nyssa's *Opera* and clearly spent some time with the volume.[9] And he was certainly interested in various understandings of the creation of the world that were prevalent among thinkers from antiquity.[10] Newton might have been interested in Gregory's claim that God's substance or nature (*phusis*) was unknown but that God's energies or operations (*energeiai*) are manifested in the world. Or he may have noted Gregory's commitment to the idea that omnipotence involves God being in, penetrating, and embracing everything. Most striking, however, are deep resonances between Gregory's discussion of creation in his *Hexaemeron* and Newton's account of creation in *De gravitatione*. Gregory's treatment focuses on the difficulty of reconciling the creation of *material* objects by an *immaterial* God with adherence to a principle according to which "like causes like", such that the features had by the effect must be had by the cause.[11] Gregory's solution involves making material substances mere collections of qualities that are essentially graspable by human minds. As will be shown later, Newton himself developed a similar view.[12] To be clear, more would need to be done to demonstrate a definite positive or negative

[7] Westfall 1980, page 51.

[8] See Mandelbrote 2006a. For some discussion of earlier engagement with these sources see Westfall 1980, pages 310–315.

[9] *NL* #702. Referred to as "Nyssen," Gregory of Nyssa is mentioned in roughly two dozen of Newton's remaining manuscripts.

[10] See, e.g., Yahuda Ms. 15.7, 137r.

[11] For a helpful reconstruction of Gregory's views along these lines see Marmodoro 2015.

[12] Robles 2001, pages 8–14, also notes a connection between Nyssa and *DG* along these lines. See also Robles 2006, page 284.

line of influence from Gregory—or any Church Father—to Newton's metaphysical views. The point is just that to the degree Newton made any use of these resources, resources which he had to hand in his formative years, he may have been exposed to rich theorizing on core metaphysical themes.

In the summer of 1661, Newton went up to Cambridge. Here, we can begin to pinpoint his intellectual activity and influences with a much higher degree of precision. As part of the standard curriculum, Newton was assigned a number of Scholastic philosophy textbooks in widespread use at the time, for example, Johannes Magirus' *Physiologiæ peripateticæ* and Daniel Stahl's *Axiomata philosophica*.[13] He seems to have worked through some of these, or at least the early chapters of them, enough to have formed a clear sense of their contents albeit with no great enthusiasm.[14] Indeed, it is hard to overlook the fact that Newton's notes on Magirus end with ch. 5 of Book IV where he copies down the proffered definition of "galaxy" before drawing a line across the page and launching into a discussion of Galileo.[15] One might see this dramatic moment as the end of Newton's philosophical career and the beginning of his astronomical one.[16]

This, however, would be a mistake. Instead, we ought to read this as the end of Newton's interest in Scholastic philosophy and the beginning of his full immersion into the philosophical world of the moderns. Nowhere is this made more clear than in the lengthy portion of a notebook kept during his undergraduate years and entitled by him *Questiones quædam Philosophicæ* (henceforth *CPQ*, for *Certain Philosophical Questions*).[17] The work shows Newton's deep reading in and careful consideration of a host of early modern philosophers including René Descartes, Robert Boyle, Henry More, Thomas Hobbes, and Walter Charleton. And it also shows his ambition to build something like a complete metaphysical system in the vein of these thinkers. This ambition was, ultimately, not realized in *CPQ*. Newton ran into various deep philosophical puzzles and came to abandon some of the positions developed in the notebook. At the same time, however, the rudiments of the system he was working toward in *CPQ* remain a part of his metaphysical worldview throughout the remainder of his life. Specifically, *CPQ* shows him finding the outlines of an idiosyncratic version of mind-body dualism, it features his abiding attraction to a broadly Epicurean conception of the material world as composed of matter and void, and it showcases his lasting desire to offer a philosophically satisfying corpuscularian (if not fully mechanical) matter theory. Lastly, the text seems

[13] On the curriculum and intellectual life of the university during Newton's undergraduate years see Costello 1958 and Part 1 of Gascoigne 1989.

[14] For details and some helpful reconstructions of Newton's notes see Wallace 1988. See also Ducheyne 2012, chs. 1.3 and 1.4.

[15] CUL Ms. Add 3996, 26v. See Wallace 1988, page 32.

[16] And Newton's first sustained astronomical reading and observations do date from shortly after this time. See McGuire and Tamny 1985.

[17] For background on and an enormously helpful guide to this work, see the introduction prepared by J.E. McGuire and Martin Tamny at *CPQ*, pages 3–325.

to reflect an enduring Newtonian approach to the relationship between God and nature. On this view, God is an agent of overwhelming power and is fully deserving of worship in virtue of being the creator of and ultimate arbiter over the universe. At the same time, Newton stops short of making God directly and immediately involved in the ordinary operations of the natural world. Subsequent chapters will explore in much greater detail Newton's engagement with his predecessors in this text, the arguments he constructs, and the positions he adopts.

The latest notes in *CPQ* date from 1665, but this fertile period of engagement with issues in metaphysics continued for some time. The highlight of this next period of reflection is *De gravitatione*.[18] This is an unfinished and untitled manuscript essay which begins with a discussion of hydrostatics before offering an extended treatment of issues relating to space, time, motion, God, creation, substance, bodies, perceptions, and minds. Much of Newton's discussion is negative and aims to dispel what he saw as the errors present in Scholastic, "common sense", and, especially, Cartesian treatments of these topics. But much is also positive, outlining and defending Newton's own views on these issues. Much of the rest of this book will be devoted to an interpretation of *De gravitatione*.

The years just canvased, the 1660s and early to mid-1670s, mark the period of Newton's deepest and most careful thinking about metaphysics. Never again, so far as I have been able to discover, did he approach these issues with the same level of investment and care. Nonetheless, the positions Newton adopted in this formative period stuck. While they underwent some revision, amendment, and development, they continued to provide the core of his metaphysical outlook until his death in 1727.[19] This, at least, is one of the claims that I argue for in subsequent chapters.

There are two channels by which we can see the older Newton continuing to hold the views developed in his late adolescence and early adult years. First, as Newton works to articulate and explore positions in natural philosophy and other fields, there are consistent echoes of *CPQ* and *De gravitatione*. Sometimes this is a result of Newton pursuing his own views in natural philosophy. For example, the manuscript text *Tempus et Locus* dates to around the time of the first edition of the *Principia* and is enormously rich in metaphysical theorizing.[20] Similarly, something very deep about Newton's fundamental understanding of the metaphysics of material substance is revealed in his correspondence with Roger Cotes on the run up to the second edition of the *Principia*. And, of course, the General Scholium added to the second edition of the *Principia* in 1713 features a number of striking

[18] For more on the background to and dating of this manuscript, as well as some of its controversial features, see Section 1.7 below.

[19] We can note a potential parallel in his theological views. As Westfall has argued, Newton's views on key theological issues were developed in the 1670s and 1680s and remained more or less fixed from that point on. See Westfall 1982, pages 138–139. Dobbs 1991, page 169, largely agrees with this assessment.

[20] McGuire 1978b. Levitin 2022, pages 802–808, challenges this dating.

metaphysical arguments and assertions. So, Newton's own career offers insights into these topics. But Newton's engagement with the thought of others also sometimes shows him grappling with themes relevant to this book. From his discussion of creation in Genesis with Thomas Burnet, to his reading of Locke's *Essay concerning Human Understanding*, to his comments on William Briggs's account of the physiology of perception, Newton's correspondence and manuscript notes offer a wealth of insight into his later metaphysical views.

The second channel that offers insight into Newton's persistent adherence to the views of his youth relates to Leibniz's efforts to engage Newton and his allies on philosophical topics and the concurrent priority dispute over the invention of the calculus. There are two common approaches to these subjects that I think we have good reason to resist. The first makes Samuel Clarke the official spokesperson of not just "Newtonianism" but also of Newton himself. This view considers the positions taken by Clarke in his correspondence with Leibniz to be Newton's own views and attempts to reconstruct his thought on the basis of them.[21] The second approach sees Newton as, at best, unwilling, or, at worst, unable to consider and respond to Leibniz's philosophical critiques. For the first, it would be obtuse to deny that a number of the positions taken by Clarke in his correspondence with Leibniz were positions that Newton himself held. But, on my view, it would be equally obtuse to claim that *all* of the positions taken by Clarke were ones held by Newton. By the time of the correspondence, Clarke was amongst the foremost metaphysicians in Britain and had articulated his own clear, coherent, and, in ways, original metaphysical system in his *Demonstration of the Being and Attributes of God*. To portray him as nothing other than Newton's mouthpiece is both misleading and unfair to both parties.[22] For the second, it is abundantly clear that Newton had no appetite for a protracted public dispute with Leibniz on philosophical issues. There are many different reasons why this was so. But one which, I believe, has not been given enough attention has to do with what Newton took Leibniz's motives to be. Specifically, Newton very often complains that Leibniz seeks to engage him on metaphysical issues in bad faith. On his understanding of the situation, Leibniz knew perfectly well that he was on the wrong side of the priority dispute. Newton claimed, therefore, that Leibniz only brought up other issues with the goal

[21] Note that the strategy being criticized here is different from the one adopted in Levitin 2022, pages 664–679 and 725–757. Levitin seeks to identify specific passages or views in Clarke's work as the proximate sources for Newton's thinking on particular issues. This is a much more targeted and historically nuanced approach. Levitin's claims on this front are intriguing, suggestive, and worthy of further consideration. That being said, my view is that in some instances the strength and scope of his conclusions outrun the evidence offered in support of them.

[22] An influential article by Koyré and Cohen is sometimes taken to show that Newton himself contributed to the content of Clarke's letters. See Koyré and Cohen 1962. But, as McMullin 1978, page 130, points out, the evidence they present is less than fully probative and even the conclusion they themselves present is highly circumscribed. As they write on page 79: "We have found no evidence to make precise either the degree of Newton's participation in the "Leibniz-Clarke correspondence" or of Clarke's participation in Newton's letters to Conti."

of providing a distraction or changing the subject. Fortunately, while Newton was unwilling to engage publicly in any sustained or detailed way, Leibniz did push him to consider issues such as the nature of atomism and the extent of divine activity in the created world. And Leibniz also pushed Newton to articulate his own views in a way that makes clear their divergence from Leibniz's.[23] In this sense, and thanks to Leibniz, we can learn about the content of Newton's metaphysical views in the final decades of his life.

This has been very brief. More will be said about nearly everything mentioned in this section in subsequent chapters of this book. Indeed, some of the texts and episodes mentioned above will be pored over in enormous detail. But my hope is that this quick survey has provided something by way of general orientation and timeline. Further, my hope is also that for some readers it will have opened up the possibility of Newton as a metaphysician by situating his engagement with metaphysical thinkers, themes, and questions against the backdrop of his life and career.

1.3. Unity and Diversity in Newton's Thought

Throughout his long career, Newton pursued a number of interests. He is obviously best-known for his contributions to mathematics and physics. But he also pursued serious work in optics, chymistry, Biblical history, chronology, and Scriptural interpretation. His professional life saw him deeply involved in economic questions, monetary policy, and legal procedure during his time at the Mint, as well as university and national politics as a result of his position at Trinity College, Cambridge, and his stint as a member of Parliament. And, most relevant for the purposes of this book, Newton gave serious thought to questions in fundamental metaphysics.

Newton's deep and sustained interest in such diverse pursuits has given rise to a series of questions about the unity or otherwise of his thought and his understanding of the world. At one extreme, and perhaps at the risk of caricature, some in the eighteenth-century were keen to see radical discontinuities in his thought and between his various pursuits. The Jeykll-like rational search for truth in terrestrial and celestial mechanics had nothing to do with the unhinged Hyde-like dabbling in alchemy. At the other extreme, a number of scholars have proposed much more synoptic interpretations. Some, like Betty Jo Teeter Dobbs and James Force, have argued that a certain understanding of God and God's immense and powerful will provides a unifying framework for Newton's various pursuits.[24] Piyo Rattansi has claimed that the search for a *prisca sapientia* is at the heart of a unified Newtonian

[23] Connolly 2024 focuses on the question of whether gravitation is an "occult quality" and can serve as an illustrative example or case study for the more general claims made here.

[24] Dobbs 1991 (a relevant selection of which has been reprinted as Dobbs 1993); Force 1990a; Force 2000. See also Snobelen 2001; Snobelen 2004; Leshem 2003.

intellectual endeavor spanning multiple domains.[25] Jed Buchwald and Mordechai Feingold, in their book on Newton's historical and chronological interests, have argued that while Newton applied himself to diverse *subject matters*, his work is nonetheless unified by the fact that he brought the same methodology and treatment of evidence to all areas of his thought.[26] By contrast, Rob Iliffe has attempted to apply a corrective to some of these deeply synoptic readings of Newton by emphasizing that—even within just the scope of natural philosophy—Newton's own understanding and presentation of his work reflected a keen awareness of various divisions between disciplinary boundaries and methodologies.[27] Thus, these questions of unity and diversity remain live ones in current scholarship.[28]

My own view, perhaps unexcitingly, falls somewhere in the middle of this spectrum. Newton was, *inter alia*, a human being. Even rather strange human beings like Newton have webs of beliefs and interests that are partly connected and partly divergent. And while some beliefs and interests will be closer to the center of the web, others will be more peripheral. It is not obvious that, say, someone's research in contemporary epistemology will be deeply connected to their desire to learn Syriac. Though one could imagine a story that would link these two tightly. It might be odd if someone embodied both a fierce commitment to and advocacy for vegetarianism as well as a love of duck hunting. But, of course, such a person is imaginable. This is just to say that, with Newton, I think we ought not have unrealistic expectations of either full systematic alignment or of deeply compartmentalized diversity. Instead, I believe the reasonable approach is a particularist one. We ought to look at things on a case-by-case basis.[29] To take a toy example, I think it is reasonable to ask whether or not Newton's conviction that there was an Athanasian conspiracy to corrupt the early Church helpfully illuminates or is helpfully illuminated by the revisions to the 'Rules for the Study of Natural Philosophy' in subsequent editions of the *Principia*. And it is reasonable to do this without antecedently deciding that the two either are or are not related in some particular way.

[25] Rattansi 1988, especially pages 191 and 193–194. Dobbs 1991 also contains a version of this claim.

[26] Buchwald and Feingold 2013.

[27] Iliffe 2004. Section 1 of Iliffe's paper offers a helpful overview of secondary literature relevant to these debates. In connection with Iliffe's claim we can note that even a "unifier" like Dobbs allowed that "Newton pursued each area of study as if it were the only one of importance to him and almost always adhered to the terminology and the concepts embedded in that particular field." Dobbs 1991, page 251.

[28] Insight into the current state-of-play on the topic is provided in the 2021 *Early Science and Medicine* symposium on "Reassessing the Wider Aspects of Newton's Thought," which takes Buchwald and Feingold 2013, Iliffe 2017, and Newman 2019 as its starting point. The contributions offer a helpful point of entry into current perspectives on the questions considered in this section. See Henry 2021; Guicciardini 2021; Mandelbrote 2021: Ducheyne 2021; Snobelen 2021; Feingold 2021; Iliffe 2021; Newman 2021.

[29] Iliffe 2004 can be read as proposing something like this. Greenham 2017 provides another nice example of this approach. Further, with some risk of overgeneralization and without wanting to paper over disagreements between the scholars involved, this sort of particularist approach is arguably the one employed (to good end) by the contributors to the "Reassessing the Wider Aspects of Newton's Thought" symposium mentioned above.

In some cases, we will uncover productive connections and linkages that enrich our understanding of Newton and his thought. In others, we will not. And in still other cases, we might expect to discover disconnects, tensions, or inconsistencies. These too can serve to enrich our understanding of Newton and his thought.

1.4. Metaphysics and Natural Philosophy

This issue of unity or diversity in Newton's thought is particularly pressing for the purposes of this book because of the various ways Newton's metaphysics has been understood in relation to his work in natural philosophy. Philosophical work on Newton is, perhaps unsurprisingly, dominated by a focus on issues in the philosophy of science. An enormous amount of work has been devoted to Newton's views on force, motion, laws, space, time, experiment, hypothesis, methodology, evidence, and the like. This is all to the good. But, often as not, when philosophers and historians of philosophy have asked about Newton's views on metaphysical questions, it has been with one eye toward these issues in science and the philosophy of science.

Consider, for instance, the helpful taxonomy of approaches to Newtonian metaphysics developed by Andrew Janiak at the outset of his landmark book, *Newton as Philosopher*.[30] First, one might understand Newton as a type of positivist. On this view, he rejects metaphysics entirely. His motivation for doing so was to insulate natural philosophy from the pernicious forces of metaphysical speculation. Second, one might see Newton as adopting a radically empiricist metaphysics. On this view, he is open to metaphysical positions and commitments but only insofar as they track scientific findings. More precisely, Newton sought to transform metaphysical questions into questions that could be settled by physical theory. On a view like this, we might end up holding metaphysical views, but only ones fully licensed by or straightforwardly derivable from solid natural philosophy. Third, one might see a more complex interplay between metaphysics and natural philosophy. There might, on the one hand, be an unrevisable metaphysical core to Newton's conception of the world, perhaps informed by his understanding of God's nature and agency. This, in turn, might provide some parameters within which natural philosophy is to be pursued. At the same time, and on the other hand, Newton might still see natural philosophy as relevant for either determining certain truths about metaphysics or as a useful tool for dispensing with misguided metaphysical inquiries.

In a very basic sense, I reject all three of these approaches. While I side most closely with the third, the one favored by Janiak, I balk at the fact it seems to

[30] Janiak 2008, ch. 2. See also Domski 2010 which offers a helpful discussion of this terrain and, in particular, of the second two possibilities outlined by Janiak.

embody a foundational presupposition shared by the other two: that what matters most for Newton's metaphysics is how it relates to his natural philosophy. To my mind, this framing embodies precisely the kind of demand for unity or systematicity in Newton's thought that, as discussed in the previous section, I think we have good reason to be hesitant about. In what follows, I will have some things to say about the links between Newton's metaphysics and his natural philosophy, especially in Chapter 6. But, in reading and thinking carefully about a number of Newtonian texts, I have become convinced that Newton had various firm and reasoned commitments in metaphysics that amount to a systematic approach to and understanding of substances and their relations. I want to examine this systematic metaphysics on its own merits: to sketch the shape of it and understand how it works. In keeping with the view articulated in the previous section, I believe it is only once a working understanding of Newton's metaphysical views is up and running that it will make sense to begin looking at those views in connection with other elements of his thought and to do the work of either fitting his metaphysics into some larger whole or allowing it to remain siloed.[31]

1.5. "Metaphysics"

I understand metaphysics to be, very roughly, systematic inquiry into the fundamental nature and components of reality and their relationship with one another: what there is and how it all hangs together. Understood in this way, traditional questions and debates in metaphysics range over topics such as God, substance, properties, persons, space, time, causation, abstract objects, and the like. This book tackles Newton's understanding of God, bodies, and minds and the relationships that obtain between these three types of substance. We will see Newton moving his way through a number of fraught debates in seventeenth-century metaphysics: the univocity of substance, the nature of the dependence relation between God and creatures, the ontology of human beings, and more. In that sense, I feel comfortable in claiming that Newton had a metaphysics or was a metaphysician.

Given this, it is worth considering the views of a recent commentator who may appear to have a different interpretation. Specifically, Dmitri Levitin has argued, in a number of works, that Newton was a staunchly anti-metaphysical thinker who had no developed or systematic metaphysics.[32] In one sense, it may be that Levitin

[31] There are some resonances between this approach and Iliffe's approach to Newton's religious thought. See Iliffe 2017, pages 22–23.

[32] See Levitin 2016, Levitin 2021, Levitin 2022. For a similar view see the Introduction in Kassler 2018. A related view, that Newton's "involvement in philosophical issues was defensive in nature" and a product of a need to respond to criticism rather than driven by a "genuine interest" can be found at Guicciardini 2018, pages 179–180. See Schliesser 2021, pages 3–8, and Brading 2025 for some sensible criticism of Levitin's anti-metaphysical interpretation of Newton. Lüthy 2025 is also relevant here. Levitin responds to Brading and Lüthy in Levitin 2025.

and I have merely a verbal disagreement. Sometimes Levitin's actual claim is that Newton is not doing a particular *kind* of metaphysics, be it Scholastic, or Morean, or a prioristic. Fair enough. But this is compatible with Newton doing some other kind of metaphysics. At other times, Levitin suggests that rather than doing *metaphysics*, Newton is doing something much more like *natural theology*. On my view, and to take just one example, an investigation of the world which singles out certain features of it for use in a design argument that issues conclusions about God and the divine nature is an exercise in metaphysics. The reason is that such an exercise will—of necessity—make or rely on deep commitments about the nature of existence, causation, modality (actuality, possibility, necessity), substance, time, providence, and intentionality. I would call this metaphysics. But if others want to call it natural theology, nothing much seems to hang on the issue.

There is, of course, a question about whether metaphysics—as I am characterizing it—is an actor's category for a figure like Newton. Here I think the issue is complicated. As Levitin and others have done so much work to establish, metaphysics was a contested discipline in Newton's period and his contemporaries had divergent and incompatible conceptions of it. Some might have included items from the list I gave above within the domain of metaphysics. Others might have seen some of these items as falling under physics, rather than metaphysics. And, of course, for some of them, physics itself was a branch of metaphysics. What about Newton? On the one hand, there are instances of Newton using "metaphysics" in the way I intend to use it. Take, for instance, this passage from *De gravitatione*:

> [T]he usefulness of the idea of body that I have described is brought out by the fact that it clearly involves the principal truths of metaphysics and thoroughly confirms and explains them. For we cannot posit bodies of this kind without at the same time positing that God exists, and has created bodies in empty space out of nothing, and that they are beings distinct from created minds, but able to be united with minds.[33]

Newton here claims that there are fundamental truths of metaphysics. In the larger context of the passage, these truths seem to include God's existence, creation *ex nihilo*, the existence of absolute space, substance dualism, and mind-body interaction. And Newton is claiming that he has done something good, namely develop an account of the nature of corporeal substance which confirms and explains these truths. Metaphysics seems to be operating as an actor's category here. And the range of topics Newton takes it to apply to—for example, substance, individuation

[33] *DG* 45. Schliesser 2021, page 4, also appeals to this claim in responding to Levitin. Levitin is, of course, aware of this passage and seeks to deflate its relevance for these issues. See Levitin 2021, page 158.

and kinds, creation, causation—lines up neatly with our contemporary understanding of the topics central to seventeenth-century metaphysics.

That having been said, there are also numerous instances of Newton using "metaphysics" and its cognates in a very different manner. This lack of linguistic rigor can make it difficult to issue sweeping claims about Newton's own understanding of "metaphysics." The right move, it seems to me, is again a particularist one. We ought to look at individual instances to determine how, in that particular instance, the word is being used by Newton, with what valence, and to what end. A helpful framework for this might be found in the sensitivity that scholarship now shows to the use of "hypothesis" and its cognates in Newton's work. Whereas Newton's famous pronouncement in the General Scholium that "hypotheses non fingo" was once taken to be a sweeping claim reflecting a global methodology, there is now widespread appreciation of the fact that Newton's use of the word varied widely across time, text, and context.[34] It is true that Newton, like many of his contemporaries, often used "metaphysics" (and, for that matter, "hypothesis") in a decidedly dismissive or pejorative sense. But I believe we need to be careful to avoid overreading these uses of the term as amounting to a categorical dismissal of metaphysics. As Eric Schliesser, drawing on an observation by Daniel Schneider, has noted in criticizing Levitin, adopting an actor's category as one's own is "often no better than to take sides in the polemics of the past without fully realizing it."[35] I think this danger of being conscripted as an unwitting foot-soldier is a very real one. It does not, of course, constitute a reason to ignore or downplay the importance of actors' categories. It just amounts to an argument for the sort of careful particularism invoked above. For that reason, I think it is certainly a possibility that metaphysics—as I am using it—is an actor's category for Newton but I believe the issue must be treated with a certain degree of sensitivity. And should it prove to be the case that it is decidedly not an actor's category, I think that would have very little significance for what happens in the body chapters of this book.

At other points, Levitin's key claim is not so much that Newton refrains from making claims or statements about metaphysical issues, but rather that it would be misguided to take these claims as amounting to or constitutive of anything like a commitment to the practice of metaphysics or the having of systematic metaphysical views. This is because, on Levitin's understanding, the claims are either unoriginal, not particularly deep or well worked-out, or are made in highly specific contexts for purely polemical purposes. Here, unlike with the use of the word "metaphysics," I think my disagreement with Levitin is substantive. In some respects, the rest of the book is an effort to show that Newton had deep commitments to carefully considered metaphysical positions which are mutually related

[34] A helpful starting point for this topic is Cohen 1966. On the linguistic flexibility of "hypothesis" in other figures from Newton's milieu, see Connolly 2026, sec. 2.

[35] Schliesser 2021, page 6.

and reinforcing in a way that allows them to be understood as a coherent whole.[36] So, with respect to this issue, the proof of the pudding will have to be in the eating.

What should we make, then, of the question about whether or not Newton was a metaphysician or had a systematic metaphysics? Consider an analogy. Was Newton interested in theology? Was he a theologian? In one sense, the answer is an obvious and resounding "no." He had no time at all for many of the abstract theological debates and doctrinal disputes that so obsessed many of his contemporaries. And there are any number of dismissive and withering remarks about theology and theologians in his corpus. But to leave matters here would paint a seriously distorted picture. Was Newton interested in religion? Did he care about and pursue fundamental questions relating to God? Did he stake out clear positions on these issues? The answer to these questions is an equally obvious and resounding "yes." In that sense, without any clear offense to either contemporary or historical notions of the term, I think we can say Newton carefully pursued theological questions and developed a theological worldview. On my view, something similar applies in the case of metaphysics.

1.6. Newton's Empirical Methodology

Newton's methodological approach to the issues addressed in this book is broadly empiricist. It is broadly empiricist in the sense that Newton believed sound reasoning on metaphysical issues ought to begin with experience and with careful attention to phenomena. Speaking about the metaphysics of body in *De gravitatione* Newton refers to "the evidences of our senses" as constituting "our sole judges in this matter."[37] Material from Newton's drafts for the Queries is even more striking: "Nor are we here [in experimental philosophy] to regard Metaphysical Principles unless so far as they are founded upon experience . . . so far as Metaphysical Propositions are founded upon experience they are part of experimental Philosophy."[38] Here again we see Newton using metaphysics approvingly and as an actor's category. But here there is an important methodological caveat. Metaphysical principles are to be welcomed, provided they are vouchsafed by empirical observation. There is no shortage of texts in which Newton excoriates his predecessors and contemporaries for beginning with or being led astray by speculation, preconceived opinions, or (using the pejorative sense of the term)

[36] In this sense I am broadly sympathetic to Janiak's claim that "Newton is systematic and philosophical without articulating a philosophical system." Janiak 2008, page 9. But I would go slightly farther than Janiak with respect to the scope and number of topics included in Newton's systematic and philosophical thought. See Janiak 2008, ch. 1.

[37] *DG* 42. See also *NCS* #40, where "we should not know Bodies for substances, were it not for their sensible qualities."

[38] CUL Ms. Add 3970.3, f. 621v.

hypotheses. And contrary to a figure like Descartes, Newton denied that we had any innate ideas, let alone a privileged class of innate ideas that would tell us about the essential natures of created substances and thereby provide a foundation for our theorizing. As Newton claims in the General Scholium, "there is no direct sense and there are no indirect reflected actions by which we know innermost substances."[39]

So Newton's view is that good reasoning begins with careful attention to observation and experiment: it must be founded on or derived from phenomena. It will, then, be worth saying something about his understanding of phenomena. Here is one clear definition from drafts for revisions to the *Principia*:

> Phenomena I call *whatever can be seen and is perceptible* whatever things can be perceived, either things external which become known by the five senses, or things internal which we contemplate in our minds by thinking. As fire is hot and water is wet, *and gold is heavy, and sun is light,* I am and I think. All these are sensible things and can be called phenomena in a wide sense; but those things are *properly called* phenomena which can be seen, but I understand the word in a wider sense.[40]

Newton, with reference to the etymology of the word, notes that a phenomenon, in a narrow sense, can just refer to something seen. But he warns us against this overly narrow understanding, professing to use the word in a wider sense. This point is important for two reasons. First, it seems to track Newton's actual usage of the term. Consider the fact that the phenomena listed, and identified as phenomena, at the outset of Book III of the *Principia* are not simple observations; they are not the kind of thing a person could "see" in a single experience. Instead, they include Kepler's laws of planetary motion. Clearly these are the product of careful reasoning and reflection made on the basis of multiple correlated observations (many of them mediated through the use of scientific instruments). Nonetheless, they can reasonably be seen as fitting under Newton's "wider" sense of phenomenon because, as he claims of the first phenomenon, they are "established from astronomical observations."[41] The second reason is it is important to keep in mind the fact that Newton is happy to employ an expansive conception of phenomena relates to the correlatively expansive scope of the claims that might be taken to be founded on phenomena. Consider this statement from one of Newton's letters to Abbé Antonio Schinella Conti:

[39] *Principia* 942.

[40] CUL Ms. Add 3965.13, f. 422v. Translation taken from McGuire 1966, pages 238–239. Italicized words reflect Newton's cancelations and interlinear insertions.

[41] *Principia* 797.

> His [Leibniz's] Harmonia præstabilita is miraculous & contradicts the daily experience of all mankind, every man finding in himse[l]f a power of seeing with his eyes & moving his body by his will. He prefers Hypotheses to Arguments of Induction drawn from experiments.[42]

Here Newton straightforwardly identifies mind-body interaction as a phenomenon; one experienced by all people at all waking moments. But, needless to say, mind-body interaction is a substantive metaphysical position.[43] The upshot of this, then, is that while Newton did commit to a kind of empiricism and wanted our thinking to be grounded in phenomena, this did not prevent him from endorsing substantive metaphysical theses. Put differently, we need not have the view that Newton's empiricist commitments automatically entail a form of positivism or phenomenalism.

Having affirmed that Newton was committed to a certain kind of methodological empiricism, I want to add two important caveats. First, as just brought out with reference to Newton's expansive understanding of phenomena, I think it is important to not misunderstand this commitment to an empirical methodology by overreading it or strengthening it such that amounts to a ban on substantive metaphysical theorizing. To deny that empiricists can do metaphysics would be overly simplistic. After all, even the most elaborate of Scholastic metaphysical systems were designed by thinkers who largely agreed that *nihil est in intellectu quod non primus fuerit in sensu*. Again, a commitment to grounding one's thinking in phenomena is compatible with constructive projects in metaphysics.

Second, I think we might not expect that every single metaphysical claim Newton is committed to will be grounded in, or, at least, obviously grounded in, observation and experimentation. Two examples will help to show that this is the case. First, as discussed in Chapters 2 and 3, Newton understood God to be a necessary being. But it is far from clear that Newton is entitled to this claim on the basis of mere empirical facts.[44] Second, Newton routinely appeals to various principles of reasoning whose empirical provenance is, at a minimum, questionable.[45] His design arguments for God's existence seem to presuppose some version of causal containment: the idea that the cause contains (in some form or other) the perfections present in the effect. Since at least Hume, the empiricist *bona fides* of this principle have been questioned. Perhaps more pressing are Newton's 'Rules for the Study of Natural Philosophy'. Clearly Newton relies on these rules (or, their close relatives in the first edition of the *Principia*) to establish many of his most

[42] *NCS* #1187. Underlining added.

[43] See also CUL Ms. Add 3970.3, f. 619r where "experience" is used to establish a number of other metaphysical theses.

[44] That said, see the effort at Stein 2002, page 298, note 19, to suggest that facts known from revelation are known through experience.

[45] For a related point, made in connection with Newton's optics, see Worrall 2000 and Hamou 2014.

important results. But they clearly embed a commitment to the simplicity or uniformity of nature, a commitment to which Newton, if he were a strict empiricist, would not be obviously entitled. Indeed, Newton, unlike, say, Descartes, faces special challenges on this front given his broadly voluntarist understanding of divine agency.

How, then, should we understand Newton's empiricist aspirations in light of his actual metaphysical practice? Perhaps this commitment to various methodological principles operates as something more like a guideline or a regulative ideal than like an iron-clad rule. By analogy, we might note that Newton devised rules for the study of Scripture and that his exegetical practice, by and large, reflects a commitment to those rules. At the same time, we should not let our recognition of the fact that Newton developed and articulated these rules and found them to be useful in his research blind us to the fact that he was at times willing to bend or break them.

1.7. Sources and Chronology

As noted above, Newton never wrote a systematic exposition of his metaphysical views. This means that they must be reconstructed from a number of different sources. These include works published during Newton's lifetime, a notebook he kept while a student, epistolary exchanges, reports from acquaintances regarding things he said or believed, and a vast trove of manuscripts of every kind left behind after his death. Importantly, these documents span several decades, from the early 1660s up through Newton's death in 1727. This means that the source material for this book presents two challenges: one relating to its heterogenous and motley nature and one relating to (potential) diachronic developments in Newton's thought. I will discuss these in turn before giving some special attention to a text known as *De gravitatione*.

In working to come to grips with and usefully deploy the diverse types of primary sources available, I have tried to be guided by two things: a general attitude of trust and an effort to find systematicity in Newton's metaphysical thought. The attitude of trust is, in its basic form, a cousin of the interpretative principle of fidelity. I have tried, where possible, to take Newton's metaphysical claims at face value and as recording his reasoned commitments. In part, this attitude has been something of a necessity for a project like this. Newton's discussions of philosophical issues are, for the most part, not particularly lengthy or sustained and they are less numerous than one might hope. So, if we are to have any hope of understanding them, we will have to be very cautious in discounting the value or relevance of any metaphysical claims he does make. Of course, there are limits to this approach. Claims made in polemical contexts need to be understood as polemical. It might be wrong to simply dismiss claims made in a student notebook as irrelevant juvenilia, but it might also be wrong to give them equal footing with claims made in multiple editions of a published text from Newton's mature years. Still, I have aimed to be inclusive and to take seriously as many of Newton's philosophical statements as possible.

The effort to find systematicity, again in its basic form, might be seen as a cousin of the interpretative principle of charity. I have tried to see Newton as endorsing a consistent and coherent metaphysical worldview. This is, in part, standard practice in reconstructing the views of historical philosophical thinkers. But it is also a product of the time I have spent working to read and understand Newton's texts. In doing so, I have come to believe that Newton gave serious thought to these issues and that it is possible to coordinate them into a systematic whole without either stretching his words or gratuitously supplying extra claims or commitments not found in the text. But doing so has required a certain amount of faith in the idea that Newton was deeply committed to certain foundational claims and that these should provide the basis for interpreting his remarks across a variety of texts and contexts.

Before moving on, it will be worth saying something about Newton's manuscript remains. Richard Westfall, writing in 1976, noted that "although there is no Newton factory, the immense quantity of Newtonian scholarship has led to talk of a Newtonian industry."[46] Even earlier, Arnold Thackray had bemoaned the fact that: "The exponentially expanding literature on Newton defies reading."[47] In the intervening half century between these claims and the present, this exponential expansion has continued apace. Surely one key driver of this has been increasing awareness of and access to Newton's manuscript remains.[48] One key step in this process was the removal of the Portsmouth Collection to Cambridge University and the publication of A. Rupert and Marie Boas Hall's *Unpublished Scientific Papers of Sir Isaac Newton*.[49] In recent years, scholars have been even more fortunate. In part, this is because of the wonderful work that Cambridge University Library, the National Library of Israel, and other institutions have done to make digitized images of Newton's manuscripts available online. But in much larger part, it is due to the extraordinary achievements of Rob Iliffe, Scott Mandelbrote, and their team at the Newton Project, who have transcribed enormous numbers of Newton manuscripts and made them freely available online. This is a tremendous service and all Newton scholars have reason to be grateful to those who have organized and contributed to the project.[50]

[46] Westfall 1976, page 175.

[47] Thackray 1970, page 8, note 1.

[48] For an account of Newton's *nachlass* and its fate following his death and up through present times, see Dry 2014.

[49] Hall and Hall 1962.

[50] Similar work has been undertaken by William Newman with the Chymistry of Isaac Newton project. I give it less attention here just because the material available on the Newton Project has been, by and large, more useful for the subjects treated in this book.

The upshot of all of this is that Newton scholars now have access to literally millions of words written by Newton. This, in turn, presents its own challenges. Even with careful and accurate diplomatic transcriptions that are text searchable, and even with a relatively narrow focus like Newton's metaphysical views, the amount of content is overwhelming. I freely confess that I have done nothing like exhaust the available material. What I have done is twofold. First, I have benefitted from and tried to build on the careful and painstaking work of scholars who have come before me and who have identified important texts and relevant passages. Second, I have worked to identify and explore some texts and passages that I believe were previously overlooked or underappreciated. They are discussed at various stages in the following chapters. I very much hope that others will benefit from their being brought to light and will be able to incorporate them into future scholarship on these topics. Put differently, it seems to me that the project of sifting through and understanding Newton's manuscript remains is so colossal that it can only be a collaborative endeavor.

Thus far this section has focused on primary sources, I now want to turn to chronology. Newton, famously, developed the calculus and discovered the theory of universal gravitation. Scholars have quite rightly been motivated to understand what led to these achievements and how they came about. As a result, many studies in the history of science and the history of philosophy of science that have Newton as their subject focus on change and on diachronic development in Newton's thought. And Mordechai Feingold has recently drawn attention to important issues of diachronic development and change in Newton's theological thought as well.[51] My strategy in this book is somewhat different; I aim to emphasize continuity in Newton's metaphysical thinking. I do not want to deny that Newton's understanding of certain metaphysical concepts did shift throughout his long career and I discuss some of these shifts in subsequent chapters. But, to my mind, the shifts were never dramatic enough to disturb the basic core of his views. In part, this foregrounding of continuity is related to the search for systematicity as a general interpretative approach discussed above. But even more so, it stems from the fact that it was in his youth that Newton gave the most attention to these issues and achieved some of his most important insights about them. It is largely the fact that he formulated the core of his metaphysical worldview early on and never saw the need to radically remake it that accounts for my decision to give relatively little weight to changes in his thought and also my willingness to draw on texts written decades apart in an effort to interpret his views.

There is one text that is most central to the interpretation developed in this paper: *De gravitatione*. This is an unpublished and incomplete manuscript essay. Historians of philosophy have long valued it for the fact that it contains some of

[51] Feingold 2019.

Newton's most detailed and most prolonged discussions of philosophical issues. Recently, however, Dmitri Levitin has cast serious doubt on the philosophical value of this text and its relevance for understanding Newton's views.[52] It is thus worth briefly considering and exploring his claims.

Ultimately, Levitin's claims fall into two camps. One is historical in nature and has to do with the origin, context, and dating of the manuscript. The other is philosophical in nature and has to do with the value and originality of the philosophical claims made in *De gravitatione* and Newton's level of commitment to them. I broadly agree with the first set of claims and broadly disagree with the second. We can consider both in turn.

Building on work done by Mordechai Feingold, Levitin's main historical claim is that *De gravitatione* is properly situated in the context of work Newton did while preparing lectures on hydrostatics at Cambridge and can be dated to the early 1670s. Levitin's findings on these matters are commendable and it is refreshing and exciting to have something close to closure on a matter that has puzzled scholars for so long.

Ultimately, my response to Levitin's more philosophical claims—that *De gravitatione* contains little by way of serious, successful, or original metaphysics—will largely echo what I said above in Section 1.4. I simply have a different assessment of the depth, quality, and fruitfulness of Newton's claims in the text. Like Howard Stein, I think it "deserves to be considered one of the most interesting metaphysical disquisitions of the seventeenth century."[53] And like J.E. McGuire, I think it contains "the systematic establishment . . . of the basic principles of Newton's own metaphysics of nature."[54] The subsequent chapters in this book are an attempt to vindicate my assessment. But I would also note that Levitin's account of the metaphysical topics canvased in *De gravitatione* is somewhat limited; his treatments focus primarily on God's relationship to the world and the ontology of space. This means they give rather short shrift to things like Newton's rich discussion of the nature of substance and, perhaps most importantly, his account of the ontology of bodies.

Levitin often warns against making *De gravitatione* a "key" used to unlock Newton's entire worldview or to interpret central concepts in the *Principia*.[55] This is fair enough. There is a genuine danger here of putting more weight on the text than it can bear. I hope I have not used *De gravitatione* as a key to Newton's entire corpus. But I allow that I have made it the backbone of the interpretation developed in this book. I have done this for two reasons. The first is that I think it is an essay of clear metaphysical depth and insight and is one of the longest and best

[52] Levitin 2016; Levitin 2021; Levitin 2022.
[53] Stein 2023, page 130.
[54] McGuire 1982, page 145.
[55] Levitin 2016, page 75; Levitin 2021, pages 134 and 160.

Newtonian texts focused on philosophical issues. The second is that I believe it can be read as cohering with and usefully illuminating and being illuminated by the metaphysical claims Newton makes elsewhere.

1.8. Outline of Chapters

At the most general level, Chapters 2 and 3 relate to God. Chapters 4, 5, and 6 focus on bodies. And Chapters 7 and 8 discuss created minds. But because I aim to offer a holistic treatment of Newton's metaphysics and aim to say something about the relationships between these three kinds of substance, there is material relevant to each kind of substance throughout the text. For example, because Newton sees God as a fundamentally different type of substance than bodies and minds, there is material about key features of created substances in Chapter 2. And because issues of divisibility and indivisibility become relevant when searching for the best characterization of Newton's dualism, there is a discussion of bodies and atomism in Chapter 7. Still, the basic division of material is as described above.

Chapter 2 begins with Newton's arguments for God's existence, focusing first on the evidence of design to be found in animal bodies and in the structure of the celestial system. A third argument, based on the fundamentally dependent nature of created substances, is given a more speculative reconstruction. This sets up discussion of a key distinction in Newton's thought between God (a fully independent substance) and created substances (which are dependent) and Newton's consequent view that "substance" is an equivocal term.

Chapter 3 engages with issues of God's nature as a necessary being and the ontology of the divine being, specifically questions about substance and attribute. Newton's understanding of what it means for God to be a necessary being leads into a discussion of his arguments that the universe gives evidence of having been the result of free providential choice, rather than of absolute necessity or random chance. While several commentators have suggested that Newton means to be engaging Spinoza on these issues, I offer reasons to be hesitant about this possibility. The chapter then asks whether God is composed of a substance and attributes or whether Newton had a more deflationary understanding, on which God is merely an attribute or collection of attributes. I reject a prominent line of interpretation in the literature which argues that Newton saw the idea of a substance (understood as a substratum or inherence base) as a philosophical error or confusion. Newton's arguments are more focused than previously thought, and are only meant to relate to *created* substances. Finally, the chapter considers the relationship between God and the divine attributes and how Newton's understanding of God as a substance with various different attributes can be compatible with his commitment to divine simplicity. I also consider the question of God's relationship to space and time and, specifically, reject the view that space and time are divine attributes.

Chapter 4 begins the discussion of corporeal substances: bodies. The starting point is a brief summary of the account of bodies developed in *De gravitatione*, an account on which God creates bodies by co-locating various powers in space. Special attention is given to the doxastic status of this account in *De gravitatione*. Most commentators see the account as highly provisional or speculative. I argue that it represents a considered commitment on Newton's part. As part of this argument, I trace the development of the view in texts that pre-date *De gravitatione* and its subsequent life in texts that post-date it. This sets up an understanding of the account of bodies in *De gravitatione* as a theory, rather than a hypothesis.

Chapter 5 begins by noting that bodies, as created substances, are fundamentally dependent on God. It then asks about the nature of that dependence relation. The dominant trend in the literature is to interpret Newton as committed to a very high level of dependence, but I argue that he understood created substances to enjoy a higher degree of independence and autonomy than is commonly recognized. The first step in getting clear on these issues involves understanding the precise relationship between the powers constituting bodies and the spaces they occupy. Once it is clear that bodies are merely located in parts of space, and do not inhere in them, there is less need for divine involvement to explain motion. Recognizing that mobility (or motivity) is one of the powers constituting bodies serves to further alleviate the need for God to be immediately involved in the operations of bodies and also helps to solve a problem relating to the unity of created substances. The chapter also considers the way in which many of the interpretations on offer are incompatible with Newton's commitment to the claim that God is not an *anima mundi*, a world soul. The outcome of this discussion is a clearer picture of the theory of bodies articulated in *De gravitatione* and an understanding that Newton is committed to a form of divine conservation, but nothing more.

Chapter 6 looks at developments in Newton's views on body in the period after *De gravitatione*. In that text, bodies were understood as being composed of three powers: impenetrability, mobility, and mind-interactivity. This presentation left the theory incomplete in two ways. First, the powers described were underspecified and Newton desired more precise, mathematical, characterization of them. Second, Newton became convinced that there were further powers (e.g., gravitation and magnetism). These needed to be identified and precisely characterized. This sets the stage for a discussion of the relationship between metaphysics and natural philosophy in Newton's thought. The chapter also considers recent proposals relating to laws and mass in Newton's thought. It pushes back against interpretations which afford laws of nature a prominent role in Newton's metaphysics of bodies. And it offers a way of linking together two Newtonian conceptions of mass which have been seen as divergent.

Chapter 7 is the first of two chapters devoted to Newton's views on created minds and the ontology of human beings. It begins with an analysis of Newton's treatment of minds in his early writings, especially *CPQ* and *De gravitatione*. I argue against the view that Newton was a monist about human ontology and against the view that he was agnostic. At least in these texts, Newton defends a version of dualism. Human minds are distinct from corporeal bodies, but like them they are constituted by collections of powers: perception, imagination, intellect, memory, and volition. Newton's claims about each of these powers, or faculties, are analyzed. This leads into a discussion of the nature of Newton's dualism. I argue that, in part because of some core features of Newton's modal metaphysics, his dualism is radically different from, and in many ways preferable to, other dualisms on offer in the period. Finally, I consider whether Newton's views on mind-body interaction are incompatible with his commitment to the third law of motion.

Chapter 8 begins with the question of whether and how Newton's views on the ontology of humans might have changed in later stages of his career. Specifically, Newton's basic understanding of created substances and his unique understanding of dualism put him in a position to offer an alternative, monist account of human nature. Ultimately, however, there is insufficient evidence to support the claim that Newton accepted such a view. Further, his invocations of the claim that space is God's sensorium provide considerable insight into his understanding of the relationship between minds and bodies and point toward a persistent dualist strand in his thought. The chapter concludes with an interpretation of these claims about the divine sensorium, claims that serve to illustrate the basic features of Newton's God, Newtonian minds, Newtonian bodies, and the ways they all relate to one another.

The conclusion begins by offering an overview of the systematic Newtonian metaphysics of substance developed in the book. It then works to articulate ways in which Newton's system is both distinctive and compelling. It does this by highlighting ways in which Newton departs from the views of his contemporaries and by noting the competitive advantages his views have over theirs. Finally, the conclusion returns to questions about systematicity and the relationship between metaphysics and natural philosophy in Newton's thought.

2
Arguments for God's Existence and God as Substance

2.1. Introduction

God is certainly the most important substance in Newton's ontology. That being said, God is also, in some ways, the substance least amenable to philosophical analysis. In the course of the General Scholium, Newton offers a methodological claim which can help us to see why this is the case:

> We have ideas of [God's] attributes, but we certainly do not know what is the substance of any thing. We see only the shapes and colors of bodies, we hear only their sounds, we touch only their external surfaces, we smell only their odors, and we taste their flavors. But there is no direct sense and there are no indirect reflected actions by which we know innermost substances; much less do we have an idea of the substance of God. We know him only by his properties and attributes and by the wisest and best construction of things and their final causes.[1]

We are only able to learn about material substances by investigating their sensible qualities: their colors, odors, flavors, and the like. We cannot know their substances or essences. With God the situation might be even worse. God has no sensible qualities: "He totally lacks any body and corporeal shape, and so he cannot be seen or heard or touched."[2] Instead we must work backward, as it were, and seek to understand God through God's actions. As Newton succinctly says in *Tempus et Locus*, "God is known from his works."[3] And while this does yield results, they are, as I will suggest below, somewhat limited.

Fortunately, at least by Newton's lights, philosophical investigation is not our only means of learning about God. Newton believed that Scripture—or rather, historically accurate and uncorrupted Scripture properly interpreted—was a reliable source of information about God's nature, actions, and providence. In this

[1] *Principia* 942.
[2] *Principia* 942. God's "invisibility" was important to Newton and features in a number of his theological manuscripts. E.g., the second of his "Twelve Articles on Religion" stresses that "The father is the invisible God whom no eye hath seen or can see, all other beings are sometimes visible." Keynes Ms. 8, 1r. See also Keynes Ms. 3, 35 and Yahuda Ms 15.4, 67r.
[3] McGuire 1978b, page 119.

Newton's Metaphysics of Substance. Patrick J. Connolly, Oxford University Press. © Patrick J. Connolly 2025.
DOI: 10.1093/9780191959240.003.0002

sense, much of Newton's approach to God was *historical* in character rather than philosophical.[4]

As discussed in the introduction, some scholars have sought to find continuity across Newton's various intellectual endeavors. And some authors have suggested his Scripture-based understanding of the deity provides a useful lens for understanding and evaluating his other efforts.[5] I am somewhat skeptical of this approach. My view is that Newton aspired to have something of a firewall between *philosophical* investigations into the Deity and Scriptural ones.[6] The first of his "Seven Statements on Religion" has it "That religion & Philosophy are to be preserved distinct. We are not to introduce divine revelations in Philosophy, nor philosophical opinions into religion."[7] Newton's desire to maintain these as distinct realms stems from his fear that philosophical reasoning and philosophical concepts could have a corrupting influence on our efforts to properly understand the true meaning of the Scriptural text. This is a consistent theme, for example, in his "Drafts on the History of Church" where the errors of those Newton deemed heretical are often attributed to their superimposition of philosophical views onto Scriptural texts or doctrines. For example, Newton writes that:

> The grand occasion of errors in the faith has been the turning of the scriptures from a moral & monarchical to a physical & metaphysical sense, & this has been done chiefly by men bred up in the metaphysical theology of the heathen Philosophers the Cabbalists & Schoolmen. For men tainted with the metaphysical principles of Philosophers Cabbalists & Schoolmen upon reading the scriptures have been apt to strain every thing to their own opinions.[8]

This type of separation between philosophical theorizing and Biblical interpretation entails the rejection of a great deal of traditional philosophical theology. Thus, as indicated above, Newton's methodological commitments put significant limits on his analysis of God as a substance.

[4] As Iliffe puts the point: "Newton's theological research focused almost exclusively on scriptural exegesis, the history of the early church, and the meaning of prophetic images." Iliffe 2017, page 6. For discussion of this aspect of Newton's theological thought see Manuel 1963 and Iliffe 2017.

[5] One version of this can be found in Buchwald and Feingold 2013. A differently inflected version of the claim can be found in Force 1990a and Ochoa 2005.

[6] For a more nuanced development of this claim and some useful distinctions, see Greenham 2017.

[7] Keynes Ms. 6, 1r. See also CUL Ms. Add 3965, 547r. Note, however, that in these texts Newton may be cautioning more against the running together of *natural* philosophy and religion than what we would call philosophy and religion.

[8] Yahuda Ms. 15.5, 97r. For similar material in "Drafts on the History of the Church," see Yahuda Ms. 15.2, 37r and 48r, Yahuda Ms. 15.3, 45r and 47v, Yahuda Ms. 15.5, 77r, 79r, 87r, 88r, and 99r, Yahuda Ms. 15.6, 110r, and Yahuda Ms. 15.7, 116r, 137r, 154r, 170r, 177v, and 183Ar. Also relevant are Keynes Ms. 3, 32, Keynes Ms. 11, 1r, and Sotheby's Lot 255.8, f. 1r (held in a private collection) cited at Snobelen 2008, page 503.

These methodological commitments should also make us very cautious about drawing tight links between certain of Newton's philosophical views and his considered theological commitments. Howard Stein, for example, has suggested that Newton's views on the nature of substance line up neatly with his anti-Trinitarianism and may even be responsible for his view that the Trinity is unintelligible.[9] Everything we know about Newton's Arianism, however, suggests that it was founded on deep convictions about Scriptural corruption and perversion of doctrine in the early Church.[10] I have little doubt that if Newton believed there was Scriptural warrant for the consubstantiality of the Father, Son, and Holy Spirit, he would have either found some way of making the theory philosophically intelligible (by adopting a suitable theory of substance or relations) or accepted the theory as true but beyond human reason's ability to comprehend.

2.2. Arguments for God's Existence: Animal Bodies

Newton was adamant that we can know of God's existence. And while he was certainly aware of the ontological argument from Descartes' presentation of it, he seemed to have little time for a priori arguments in favor of God's existence.[11] His view was that "there is no way without revelation to come to the knowledge of a Deity but by the frame of Nature"[12] and consequently "in proving a Deity all arguments not taken from Phænomena are little better then [sic] dreams."[13] As mentioned above, Newton denied that we know God's essence. So it is perhaps unsurprising that he did not embrace the ontological argument and preferred a posteriori arguments.

In keeping with this, much of Newton's thinking on this front is teleological in nature. A passage from "Of the Religion of the Iews & Christians" offers a concise statement of the general approach: "The wisdom & power which appears in the frame of the world & of its various parts is sufficient to convince men that they were framed by a wise & powerful being."[14] This much is relatively unremarkable. What

[9] Stein 2002, page 282.

[10] See Iliffe 2017. See also Wojcik 2000, pages 186–191 which helpfully contrasts Newton and Boyle on this point.

[11] For his notes on Descartes' ontological argument see *CPQ* 465. For discussion of Newton's principled opposition to the ontological argument see McGuire 1978a, especially sec. 4. Descartes' ontological argument also receives a brief treatment at CUL Ms. Add 9597.2.14, 4r. See Ducheyne 2012, page 260.

[12] Yahuda Ms. 41, 6r.

[13] CUL Ms. Add 3970.3, 622r intended for 621v. See also CUL Ms. Add 3970.3, 619r.

[14] "Of the Religion of the Iews & Christians," 1r. The location of this manuscript is currently unknown. It was part of Lot 255 in the 1934 Sotheby's sale and part of Lot 511 in the 2004 Sotheby's sale, where it was not sold. This manuscript also contains a canceled section that offers some brief non-teleological arguments against the eternity of the universe, thereby supporting a belief in creation in time.

is perhaps more interesting is that Newton, while his natural philosophical efforts were primarily directed at physics, astronomy, and chymistry, was initially most attracted to teleological arguments drawn from the life sciences. Specifically, he believed that the structure and organization of animals provided decisive evidence that they had been designed by an intelligent agent.[15]

CPQ offers an early version of this approach to the teleological argument. Under the heading "Of God" he writes the following:

> Were men and beasts made by fortuitous jumblings of atoms there would be many useless parts in them, here a lump of flesh, there a member too much. Some kinds of beasts might have but one eye, some more than two, and others two eyes.[16]

There are three things worth emphasizing in this passage. First, the general structure of the argument. We can know that animals were designed and created by God for two reasons: (1) their bodies are orderly and constructed parsimoniously, and (2) there are basic similarities that obtain between many different types of animals (e.g., that they have two eyes). Second, Newton targets his argument against a version of Epicureanism. The suggestion is that the only viable alternative to creation by an intelligent agent is that the present frame of things arose through random chance. But random chance is insufficient to account for the relevant phenomena.[17] Third, Newton's version of the argument has a particular focus on the eyes of animals and the fact that animals (tend to) have two eyes. These three features of the passage are notable because they are distinctive of Newton's thinking on the issue throughout his career.

All three are present, for example, in a draft of the "Hypothesis concerning Light and Colors," which repeats the content of the argument. Again Newton affirms that the "best" argument for God's existence is "the frame of nature & chiefly the contrivance of the bodies of living creatures."[18] And again the eye receives detailed attention, though this time as one representative of the many finely calibrated components of animals. Notable in this presentation of the argument is an elaboration on the way in which the similarities between animals show that different species could not have been the product of random chance. After noting that "all the great land animals" are alike in having two eyes, a nose between the eyes and mouth, two ears on the side of the head, and so forth, he argues that "this symmetry in the

[15] Guerlac 1983 discusses biological thought in Newton's design arguments and identifies some potential sources of his thinking and the examples he gives. Levitin 2022, pages 669–671, identifies correlates in Clarke's work.

[16] *CPQ* 447. The claim about animals lacking "useless" parts aligns neatly with Newton's claim at *NCS* #264 that "God in ye frame of animals has done nothing without reason."

[17] In connection with this we can note that, long after writing this note, Newton dog-eared a page in his copy of Locke's *Essay* that contained arguments against this kind of Epicurean reasoning. See *NL* #967 and Locke 1690, page 359.

[18] CUL Ms. Add 3970.3, 479r.

several species could not proceed from chance there being an equal chance for one eye or for three or four eyes as for two, & so of the other members."[19]

Other texts also show Newton's adherence to this general line of thought. In a "Treatise on Revelation" from the mid- to late 1680s he wrote that "He that shall well consider the strange & wonderful nature of life & the frame of Animals will think nothing beyond the possibility of Nature, nothing too hard for the omnipotent power of God."[20] And in "Irenicum, or Ecclesiastical Polyty tending to Peace," written after 1710, God is characterized as "A being whose wisdom & power is manifested in the contrivance & frame of the world & the things therein & particularly in that of the bodies of animals."[21] Another text written after 1710, the "Short Schem of True Religion," contains an expanded form of this argument under the heading "Of Atheism":

> Can it be by accident that all birds beasts & men have their right side & left side alike shaped (except in their bowells) & just two eyes & no more on either side the face & just two ears on either side the head & a nose with two holes & no more between the eyes & one mouth under the nose & either two fore leggs or two wings or two arms on the sholders & two leggs on the hipps one on either side & no more? Whence arises this uniformity in all their outward shapes but from the counsel & contrivance of an Author? Whence is it that the eyes of all sorts of living creatures are transparent to the very bottom & the only transparent members in the body, having on the outside an hard transparent skin, & within transparent juyces with a crystalline Lens in the middle & a pupil before the Lens all of them so truly shaped & fitted for vision, that no Artist can mend them? Did blind chance know that there was light & what was its refraction & fit the eys of all creatures after the most curious manner to make use of it? These & such like considerations always have & ever will prevail with man kind to beleive that there is a being who made all things & has all things in his power & who is therfore to be feared.[22]

We have here again the general shape of the argument. The basic idea is that the order and structure of animals are sufficient to confute the atheist. Again, the argument is targeted at the Epicurean idea that these things could have come about randomly; Newton affirms that these things could not be "by accident" or the product of "blind chance." And, finally, there is a special focus on the eye.

Operating as we do in a post-Darwinian context, it can be difficult to get a sense for how Newton's contemporaries, whether proponents of teleological arguments

[19] CUL Ms. Add. 3970.3, 479r.
[20] Yahuda Ms. 9.2, 140r.
[21] Keynes Ms. 3, 35.
[22] Keynes Ms. 7, 1r.

or committed Epicureans, might have responded to these sorts of considerations. What would be the likely range of numbers of eyes for animals created by chance? Zero to four? One to infinitely many? And what kind of distribution might be expected over this range? The answers to these questions are, I presume, far from clear. Questions about how to responsibly calibrate our views on the possible explanations for the universe and its various features, and how to update those views make design arguments intrinsically difficult to evaluate.[23] Leibniz is one early modern author who came close to recognizing the difficulties surrounding these issues. One of his dialogue characters remarks that, if there truly are infinite possibilities then, among all worlds "there should necessarily be also some well-made ones."[24] It follows then that some agents like us will be in those worlds and if we are those agents then "there is no reason to be marveled, therefore, if it turns out that we find ourselves precisely in a world of barely satisfactory beauty."[25] The orthodox counterpart in the dialogue is forced to allow that, strictly speaking, this is true. But he then presses on with an emphasis on the improbability of us being in such a world, thereby arguably missing the point and certainly leaving the issue unsettled.

The passages introduced thus far provide valuable context for what is likely Newton's best-known statement on these issues, the one from Query 31 of the *Opticks*.[26] Here again the general strategy is to argue that the anatomy of animals provides evidence of design. But the argument as presented in the Queries offers perhaps the clearest differentiation between two distinct strands of thought that are both present in earlier formulations of the argument but sometimes run together. First, Newton emphasizes the similarities that obtain between different species of animals: the fact that they are (more or less) symmetrical and that certain structural similarities obtain between, for example, the placement of their sense organs. This fact provides evidence for God's existence because it indicates that deliberate *choice*, rather than random chance, must have been in play when the species were created. Put differently, this strand of the argument relates to God's *will*. Second, Newton points to the "contrivance" of the "very artificial" parts of animals.[27] Here the focus is on the ways in which specific parts of animals (e.g., eyes, larynxes, swimming bladders, etc.), because so complex and so well-suited to serve the needs of the animal in the context of its environment, must have been designed by a wise, skillful, and powerful agent. Put differently, this strand of the

[23] For helpful discussion of this point in the context of early modern philosophy see Sober 2023.

[24] "Conversation du Marquis de Pianese et du Pere Emery" in Leibniz 2008, page 184.

[25] Leibniz 2008, page 185. This foreshadows the so-called anthropocentric objection to contemporary fine-tuning versions of the teleological argument.

[26] The first, 1704, edition of the *Opticks* contained sixteen Queries. In the Latin *Optice* of 1706 another seven were added and numbered 17–23. The second English edition of 1730 added a further eight, which were placed in between those from the 1704 *Opticks* and those from the 1706 *Optice*. Thus, the 1706 *Optice* Queries were renumbered 24–31. In this book, I refer to Queries using the number they bear in the 1730 edition of the *Opticks*.

[27] *Opticks* 378.

argument relates to God's *intellect.* This separating out of the relevant phenomena into two distinct classes and the accompanying suggestion that the two classes provide different types of evidence for divine creation represents a clear improvement in the argument.

2.3. Arguments for God's Existence: Celestial Phenomena

There are other significant differences between the design argument presented in the Queries and the other versions discussed above. One key difference is the fact that the anatomy of animals shares center stage with a second set of phenomena drawn from the realm of astronomy. Specifically, the argument emphasizes the "wonderful Uniformity in the Planetary System."[28] It is clear that Newton takes the uniformity to consist in the fact that all of the planets orbit in the same direction and in "concentric" orbits, but it is not entirely clear what he takes this concentricity to consist in. He could be referring to the fact that the smaller orbits are wholly contained within the larger ones, or to the fact that the orbits are all (roughly) on the same plane, or to both. In any event, the thrust of the argument is clear; if the solar system had come about by chance we might expect the orbits to be far more eccentric and some to proceed in a clockwise direction while others proceeded in a counter-clockwise direction (as viewed from the Earth's north pole). But the uniformity we observe must have been a matter of (divine) choice.

A second key difference relates to the fact that the target of the argument in the Queries seems to have changed. The other arguments discussed above were cast in opposition to the Epicurean idea that the structure of the universe could be the result of pure chance. But here is what Newton says in the Queries:

> For it became him who created them [the particles of matter] to set them in order. And if he did so, it's unphilosophical to seek for any other Origin of the World, or to pretend that it might arise out of a Chaos by the mere Laws of Nature.[29]

The emphasis on arising out of chaos might be broadly compatible with a critique of Epicurean views. And following just on from the quoted passage there is the suggestion that "blind fate" is explanatorily inadequate to explain the roughly uniform nature of the planetary orbits. But the quoted passage contains two important shifts. First, Newton wants to emphasize that God not only created material substances, but also *ordered* or *organized* them. Second, the point is not just that chaos or randomness could not produce the universe that we see, but rather that chaos combined with the laws of nature could not do so. These shifts suggest that Newton

[28] *Opticks* 378.
[29] *Opticks* 378.

sees his teleological argument in the Queries as targeting Cartesian approaches to divine creation and the formation of the universe.[30] The dog-earing in Newton's copies of Descartes' *Discourse on Method* and *Principles* shows that this topic was one, admittedly of many, that piqued his interest while reading.[31] And this passage should serve to show that Descartes fits the bill as someone who held that God had created matter but that nothing more than chaos and the laws of nature was needed to bring about the current structure of the universe:

> I therefore supposed that God now created, somewhere in imaginary spaces, enough matter to compose such a [new] world; that he variously and randomly agitated the different parts of this matter so as to form a chaos . . . and that he then did nothing but lend his regular concurrence to nature, leaving it to act according to the laws he established.[32]

It is also notable that Newton, in the target passage from the Queries, uses language quite similar to that used by John Keill, in a book owned by Newton, to criticize some views of Thomas Burnet which were partially inspired by Descartes.[33] Unfortunately, while Cartesianism does seem to be the clear target, Newton's reasoning here is tightly compressed. Obviously, the claim is that even if the laws of nature operating on chaos could bring about the orbits of the planets, they would not bring about orbits as regular as the ones we see and orbits that all move in the same direction. But, beyond this there is no detailed engagement with the specifics of Descartes' views.

Newton's decision to do little more than gesture at a teleological argument based on the structure of the solar system might have had more to do with what he felt was appropriate at this stage in the Queries than with any inability to supply a more detailed argument. The fact is that by the time the Queries were drafted he had devoted considerable thought to this kind of argument and how it might go. He was prompted to do so by his exchange with Richard Bentley in late 1692 and early 1693.[34] Bentley had been selected to give the inaugural set of Boyle Lectures. The lectures, as stipulated in Boyle's will, were oriented toward "proveing the Christian Religion agt notorious Infidels, viz. Atheists, Theists, Pagans, Jews

[30] That being said, there are some important links between Cartesian and Epicurean cosmogony. For discussion see Wilson 2008, ch. 3.3. See also CUL Ms. Add 9597.2.14, 4r, transcribed and translated at Ducheyne 2012, page 260.

[31] For details see *CPQ* 144–145. Descartes' fullest exposition of these issues occurs in *Le Monde*. But, likely in large part because of the complicated publication history of this text, Newton's access to these Cartesian views came through the *Discourse* and *Principles*. Davis 1991, pages 10–11, mentions the quoted passage from the Queries in connection with Descartes, singling out *Le Monde*.

[32] Discourse on Method, Part 5. *CSM*, I, 132. See also 133–134. For the closely related text from *Le Monde*, see *CSM*, I, 90–91. For the closely related text from the *Principles*, see *CSM*, I, 257.

[33] Keill 1699, page 11. *NL* #876. This point is very briefly made at Levitin 2022, pages 670 and 679, which also notes similar moves made by Clarke.

[34] For some discussion and background, see Kerszberg 1986 and Hoskin 1988.

and Mahometans, not descending lower to any Controversies that are among Christians themselves . . . "[35] One way Bentley sought to establish the principles of religion was through an appeal to findings from recent natural philosophy, including the findings of Newton's *Principia*.[36]

Most of Bentley's side of the correspondence, including the initial letter to Newton that began their exchange, is now lost. But Newton's letters to Bentley show him developing a careful teleological argument for God's existence. It is intriguing, in part, because of the way it parallels the two-part structure of the teleological argument based on animal physiology in the Queries. Recall that one strand of that argument focused on God's will. Why should a great variety of species all be constructed in the same way rather than many different ways? If we posit an agent who *chose* to make them all one way rather than another, rather than random chance as the cause, we can explain this. And recall that the other strand focused on God's intellect. Some features of animals are so delicate and complex that only a wise contriver could have fashioned them. Considerations of both these types are present in Newton's letters to Bentley.

First, there are arguments relating to evidence of free choice in the structure of the solar system. The sun is a hot and luminous body and this differentiates it from the planets. A "blind" cause "wthout contrivance & designe" would have produced bodies all of the same type.[37] And Descartes' explanation for the difference between luminous and non-luminous bodies is inadequate. This gets us that only a free choice could explain the sun's presence and uniqueness: "Why there is one body in our Systeme qualified to give light & heat to all ye rest I know no reason but because the author of ye Systeme thought it convenient, & why there is but one body of this kind I know no reason but because one was sufficient to warm & enlighten all the rest."[38] Newton's claims here are particularly interesting in light of Book 3, Proposition 8, Corollary 5 in the first edition of the *Principia*. There he had written that "God placed the planets at different distances from the Sun so that each one might, according to the degree of its density, enjoy a greater or smaller amount of heat from the Sun."[39] I think we might see Newton's comments to Bentley as an elaboration on this short and somewhat cryptic reference to divine design in the original text of the *Principia*.

The second stage of the argument relating to free will more closely foreshadows the argument of the Queries. Just as animals are all similarly structured, the orbits of the planets are similarly structured. They all move "the same way & in the same

[35] Boyle's will is reproduced as Appendix I in Maddison 1969. This passage occurs on page 274.
[36] See Connolly 2017.
[37] *NCS* #398.
[38] *NCS* #398.
[39] *Principia* 814. For discussion of this passage, including speculation about Newton's reasons for removing it from subsequent editions, see Cohen 1969, pages 525–532.

plane without any considerable variation."[40] But the fact that comets have orbits which are hugely erratic shows that "there is no natural cause wch could determine" all the planets to have roughly similar orbits. Put differently, if the solar system were the result of chance or purely natural causes then the orbits of the planets and their moons would be as eccentric as those of the comets. Because they are regular, we can infer that an intelligent agent chose for them to be the same and to look as they do. It must have been "the effect of Counsel."[41] So the Bentley correspondence shows Newton appealing to divine will and choice in the context of design arguments.

Second, there is an argument relating to divine wisdom in the structure of the solar system. The argument here is that the orbits of the planets are delicate and very carefully calibrated in a manner similar to the ways in which certain parts of animals, their eyes for example, are delicate and carefully calibrated.[42] A planet's orbit around the sun, if it is to be stable, requires the planet's mass, distance from the sun, and velocity to all be suitably proportionate to one another. This means that the cause of the system must have known the relevant quantities, distances, and forces and must also have been able to compute suitable orbits. As Newton puts the point: "to compare and adjust all these things together in so great a variety of bodies argues that cause to be not blind & fortuitous, but very well skilled in Mechanicks & Geometry."[43] Subsequent letters to Bentley contain a further argument to the effect that the forces observed in nature, while sufficient to sustain the planets in their orbits, are insufficient to explain the origin of only slightly elliptical orbits like those of the planets. So again, "ye divine arm" is a necessary explanatory posit.[44]

This dual emphasis on God's will and intellect is again on display in a version of the argument from celestial phenomena present in the General Scholium: "This most elegant system of the sun, planets, and comets could not have arisen without the design and dominion of an intelligent and powerful being."[45] Design and intelligence speak to the wisdom needed to construct the planetary system, and dominion and power speak the role of God's omnipotent will in that construction.

Chris Smeenk and Eric Schliesser have offered a reconstruction of this part of the General Scholium that identifies a number of phenomena as key pieces of evidence for God's existence, wisdom, and power. They specifically pick out that the orbits of celestial bodies are (1) law-governed, (2) hinder each other minimally, and (3) are jointly beautiful before then pointing to the fact that (4) there is

[40] *NCS* #398. In the original version of Book III of the *Principia*, the *Treatise of the System of the World*, Newton refers to this phenomenon as a "law of the Planets" (*lex Planetarum*). See Newton 1728, page 91 (English translation at Newton 1731, page 129).
[41] *NCS* #398.
[42] For a helpful reconstruction of this aspect of the correspondence, see Koyré 1965, ch. 4.
[43] *NCS* #398.
[44] *NCS* #399. The argument is briefly reiterated in *NCS* #403 and further clarified in *NCS* #406.
[45] *Principia* 940.

immense distance between the various solar systems to ensure they remain separate and (5) the motions of the comets through the system are such that they offer minimal disruption of the system's gravitational equilibrium.[46]

While some of these phenomena are given a prominent place in this section of the General Scholium, I am disinclined to assign them a central role in the argument in the way that Smeenk and Schliesser do. The key claim for current purposes is as follows: "This most elegant system of the sun, planets, and comets could not have arisen without the design and dominion of an intelligent and powerful being."[47] I take this to be the *conclusion* supported by considerations offered earlier in the text. This gives us some reason to discount the fourth consideration mentioned above, the one about the distances between fixed stars, which occurs *after* the key sentence just quoted and which was only added to the third edition of the *Principia* and thus was missing from the original text of the General Scholium. Newton there seems to be deducing consequences of God's existence and providence rather than marshalling new evidence.

What should we make of the first consideration, that the orbits are law-governed? Newton's explicit mentions of laws occur not in the paragraph with the key conclusion about divine design but rather in the previous two paragraphs. Thus I think we are better off reading these considerations as part of the criticism of vortical theory that opens the General Scholium rather than as a part of the argument from design. Indeed, as Smeenk and Schliesser themselves note, if Newton were appealing to the fact that the orbits are law-governed as evidence of divine providence then his argument would be question-begging against certain forms of necessitarianism.[48] Newton's point is just that planets and comets obey the same laws and that this is inconsistent with the vortical theory. As for the second phenomenon, I do not see Newton make special reference to the fact that the orbits of the planets do not hinder one another beyond his comments about comets spending most of their orbits in aphelia, which was the fifth phenomenon. As for the third, it is not obvious that Newton makes any clear assertion about the beauty of the orbits. He thinks they are regular and he refers to the system as most elegant, but it is hard to see these as independent or free-standing considerations that would support the kind of argument he is attempting to make. What Newton would need to do is identify the features of the system that make it beautiful or elegant.

Here is an alternative interpretation of the argument in this part of the General Scholium, one that highlights rather different phenomena. The initial impetus for this reconstruction comes from points Newton had made to Bentley: that the planets have structurally similar orbits whereas the comets have eccentric ones and

[46] Smeenk and Schliesser 2013, pages 145–146. See the similar analysis at Schliesser 2012, page 454. Note that Mary Domski has now agreed to be listed as a coauthor of Schliesser 2012: Schliesser 2021, "A Note on the Chapters."

[47] *Principia* 940.

[48] Smeenk and Schliesser 2013, page 146.

that a delicate balance of forces is needed to produce stable orbits. Newton had thought more about these points while composing revisions for the second edition of the *Principia*, to which the General Scholium was first added. A manuscript with proposed alterations to Book 3, Proposition 41 has it that God:

> by his own counsel and will disposes (through Ministers) all things in the best order, by removing fixed stars to convenient distances lest they fall into one another, by placing the Planets in concentric orbs in the same plane with conspiring motions, and the comets in excentric orbs and various planes with contrary motions, and by granting to each one its just determination of motion and velocity by which they may be able to describe their designated orbs.[49]

Here Newton does mention the distances between fixed stars, but his main focus is on the features of celestial orbits picked out in the Bentley correspondence.

Sure enough, these features are also mentioned in the General Scholium. The paragraph which contains the conclusion about our elegant system needing to have been designed starts as follows: "The six primary planets revolve about the sun in circles concentric with the sun, with the same direction of motion, and very nearly in the same plane."[50] He then notes that the secondary planets of the solar system have concentric orbits, which move in the same direction, on roughly the same plane. We then get a contrast with comets, which move eccentrically. These, I think, are the relevant features of the elegant system. We know it could not have arisen from necessity, otherwise all orbits would either be random or would be the same. The fact that some are the same whereas others are radically different gives evidence of choice and providence.

Further support, albeit indirect support, for this interpretation comes from William Derham's recollection of a conversation he had with Newton, a recollection he sent to John Conduitt in 1733. Derham mentions a "peculiar sort of Proof of God" that Newton shared with him in conversation shortly after the publication of his *Astro-Theology*. This book was first published in 1715, so the conversation must have taken place within a few years of Newton formulating the key argument in the General Scholium. According to Derham: "He [Newton] said there were 3 things in the Motions of the Heavenly Bodies, that were plain Evidences of Omnipotence & wise Counsel: 1. That the Motion imprest upon these Globes was Lateral, or in a Direction perpendicular to their Radij, not along them or parallel with them. 2. That the motions of them tend the same way. 3 That their Orbits have all the same, or nearly the same Inclination."[51] This tracks rather closely the parts

[49] CUL Ms. Add 3965, 152v. This translation is taken from Cohen 1969, page 531.

[50] *Principia* 940. That these texts occur in the same paragraph is obscured in the Cohen and Whitman translation, which introduces a paragraph break not present in the Latin text.

[51] Keynes Ms. 133, 10–11.

of the General Scholium text highlighted above. Therefore, I think we ought to see these considerations, rather than those highlighted by Smeenk and Schliesser, as the engine of the argument.

Newton lived during the beginning of a golden age for British physico-theology. A host of British authors in the late seventeenth-century and early eighteenth-century, many of them inspired at least in part by Newton's findings, sought to marshal evidence from natural philosophy in order to establish God's existence and providence. During Newton's lifetime, several notable efforts on this front came from authors like John Ray, Richard Bentley, and William Derham. And after his death the tradition continued apace, reaching a high point in authors like William Paley.[52] Viewed alongside the works of these authors, Newton's reflections might appear to be somewhat underwhelming. It was only under prompting from Bentley that he began to formulate anything like a detailed argument. Why might this have been?

I can see two reasons that Newton might have rested content with only sketchy versions of these teleological arguments. The first is that he might have found the general argument to be so overwhelmingly obvious and so clearly compelling that little was needed by way of actual discussion or elaboration. His confidence in this type of argument is reflected in the following two passages:

> Atheism is so seseless [senseless] & odious to mankind that it never had many professors.[53]

> These & such like considerations are the most convincing arguments for such a being & have convinced mankind in all ages that the world & all the species of things therein were originally framed by his power & wisdom. And to lay aside this argument is unphilosophical.[54]

We saw a similar sentiment in one of the block quotations above, where Newton claimed that design arguments "always have & ever will prevail with man kind" in questions surrounding God's existence. An argument that commands universal or near-universal assent is not one that needs detailed development.[55]

[52] For an account of design arguments in the early modern period that foregrounds Newton, see Hurlbutt 1965. More generally see Calloway 2014.

[53] Keynes Ms. 7, 1r.

[54] CUL Ms. Add. 3970.3, 479r. A previous paper of mine, Connolly 2024, contained a more sustained discussion of this text. I would like to take this opportunity to correct an unfortunate mistake I made in that discussion. I claimed there that the text was written in 1675. I now believe the text to date from much later, from the early 1700s. On this dating, see McGuire 1970. I regret the error and am hopeful that it is not too injurious to the overall argument of my earlier paper.

[55] A similar confidence in this type of argument can be found in Locke. See Locke 1975, pages 89 and 94–95.

A second, more interesting, reason that Newton may have felt no need to spell out detailed teleological arguments is that he believed there was a simpler and more straightforward route toward establishing God's existence.

2.4. A "Direct" Argument for God's Existence

I claimed at the outset of this chapter that God is the most important substance in Newton's metaphysics. The reason for this is that all other substances depend on God for their existence. So God is the first, or primary, substance in Newton's ontology. While Newton never clearly elaborates it as a formal argument, he certainly believed that the dependence of other substances on God furnished us with further evidence for God's existence.

We can begin to get some sense for this by returning to the Bentley correspondence. Newton's first reply to Bentley begins with the following claim: "When I wrote my treatise about our Systeme I had an eye upon such Principles as might work wth considering men for the beleife of a Deity & nothing can rejoice me more then to find it usefull for that purpose."[56] On the one hand, this is unremarkable. Newton was deeply religious and clearly did see links between his theological commitments and his work in natural philosophy. And Bentley's initial letter likely informed Newton of his general aim to use Newtonian natural philosophy as part of an argument for God's existence. Newton would have no doubt welcomed this, especially given Bentley's prominent status, the high profile of the Boyle Lectures, and the fact that even at this early stage there were worries about the heterodoxy of the worldview outlined in the *Principia*.[57] But it is not at all obvious from this claim that the "principles" Newton had in mind relate to the teleological arguments that were discussed above and that occupy so much of their correspondence.[58] For one thing, Newton reports that he is happy to *find* that his natural philosophy is good for theological purposes, suggesting that Bentley's letter might have led him to seriously consider these specific sorts of arguments for the first time. This impression is strengthened by one of his claims in a subsequent letter to Bentley: "The Hypothesis of deriving the frame of the world by mechanical principles from matter eavenly spread through ye heavens being inconsistent wth my systeme, I had considered it very little before your letters put me upon it."[59] There is an important caveat here about matter evenly spread through the heavens—an important topic in the correspondence—but the general claim is that Newton had

[56] *NCS* #398.

[57] For some relevant discussion of the Epicurean threat facing Newton, see Schliesser 2013a; Guicciardini 2018, pages 199 and 206–207; Wilson 2019a. On Epicureanism and its perceived impious nature in the period, see Wilson 2008.

[58] On the concept of principles in Newton's thought, see McGuire 1970 and Walsh 2017.

[59] *NCS* #403.

not devoted much attention to the sorts of issues at the heart of these teleological arguments.

What, then, were the principles that Newton had his eye on when he set to work on the *Principia*? One hint comes from a cryptic remark at the end of Newton's first letter: "There is yet another argument for a Deity wch I take to be a very strong one, but till ye principles on wch tis grounded be better received I think it more advisable to let it sleep."[60] Richard Westfall thinks it likely that Newton had in mind "the argument from the providential course of history as foretold in the prophecies."[61] And James Force has sought to flesh out and defend this suggestion.[62] This is certainly one possibility. But I am somewhat skeptical that this is what Newton had in mind here. Newton certainly did believe that Biblical prophecies had been fulfilled and that this reflected God's special providence acting in history. But so did nearly everyone else in Britain at that time. Admittedly, Newton disagreed with others over the details of this view and over the proper method of Biblical interpretation. And certainly he had more expertise with regard to the prophetic books of Scripture than many or most of his contemporaries. But I find it hard to think that Newton would not feel comfortable sharing, at least in vague outline, the idea that fulfilled prophecies provide evidence of God's existence with Bentley. It does not seem like this is an argument whose "principles" were not well received in the period. Instead, I think we should consider the possibility that Newton had in mind the theory of bodies developed in *De gravitatione*.

We can start by noticing that Newton's caginess here bears a resemblance to Locke's caginess in a similarly cryptic passage from Book IV, ch. 10, of the *Essay*. The focus there is on creation *ex nihilo* and how it might be possible. He claims that "if we would emancipate ourselves from vulgar notions" we might begin to understand "how matter might at first be made, and begin to exist, by the power of that eternal first Being." But a full discussion of this, according to Locke, "would perhaps lead us too far from the notions on which the philosophy now in the world is built."[63] Locke comes close to saying that the principles on which the relevant account is grounded are not at present well received and it will be best to let the matter sleep. Pierre Coste tells us in a footnote that Locke's reflections in this part of the chapter were indebted to Newton. In particular, it transpires that Newton had, in conversation, conveyed to Locke and the Earl of Pembroke a version of

[60] *NCS* #398.

[61] Westfall 1980, page 505. An alternative, to my mind somewhat enigmatic, suggestion is made at McGuire 2000, page 287.

[62] Force 1990b.

[63] Locke 1975, pages 628–629. A further link between this part of the *Essay* and Newton's views on God comes from a point about language. Locke writes that the investigation could only proceed "so far as Grammar it self would authorize." There is a resonance here with Newton's point about grammar ("unavoidable narrowness of language") being an obstacle to a clear understanding of the relationship between space and God in drafts of his preface to the Des Maizeaux edition of the Leibniz–Clarke correspondence. See Koyré and Cohen 1962, page 96–101.

the account of creation similar to the one in *De gravitatione*.[64] The principles undergirding that account of creation, while not widely shared, might be seen as constituting a Newtonian argument for God's existence.

The general suggestion regarding the creation of bodies in *De gravitatione* is that they might be nothing more than certain powers that God had located in space. Newton sees any number of advantages to this way of thinking about material substances and their origin. But a key one, discussed at the end of the account, is theological in nature. Newton writes that this conception of body is "useful" in part because "we cannot posit bodies of this kind without at the same time positing that God exists, and has created bodies in empty space out of nothing."[65] Newton next offers a contrast with the views of Descartes that helps to illuminate his thinking.[66] According to Descartes, bodies just are extension. But extension is uncreated and eternal. This means that it is possible to think of bodies while imagining God not to exist (and therefore to have not created them). This, according to Newton, is a path to atheism.[67] The contrast is meant to be clear. A proper understanding of bodies, one along the lines Newton develops in *De gravitatione*, makes it clear that bodies could not exist without God having created them. Thus, the very idea of body contains the idea of dependence on God: "For certainly whatever cannot exist independently of God cannot be truly understood independently of the idea of God."[68] These then, are principles that provide a firm foundation for the belief in God. The mere existence of bodies, no matter how organized or structured or similar to one another, is sufficient to turn the mind to God's existence and creative power.[69]

2.5. Primary and Secondary Substances

The previous section outlined the way in which the dependence of all other substances on God for their existence can provide a simple or direct argument for God's existence. A proper understanding of the objects we regularly see and engage with will involve an understanding of God as their creator. Their existence speaks to God's existence. But this recognition of the dependence of all other substances on God is also important for the way in which it marks a categorical difference between different types of substances in Newton's ontology.

[64] For more on this point, see Chapter 4, Section 4.6.

[65] *DG* 45.

[66] Shein 2012 provides a helpful discussion of this anti-Cartesian argument in *DG*.

[67] According to John Craig, an acquaintance of Newton and one of David Gregory's students, Newton was even convinced that Descartes had deliberately constructed his philosophy as a "foundation of infidelity." See Keynes Ms. 132, 3.

[68] *DG* 47.

[69] A version of this line of thought can also be found in Charleton and in Barrow. See Charleton 1652, page 39, and Barrow 1734, pages 109–110.

Some helpful background for this can be found in Descartes' *Principles*. *Principles*, Part 1, sec. 51 offers a definition of substance: "By *substance* we can understand nothing other than a thing which exists in such a way as to depend on no other thing for its existence."[70] As Descartes goes on to note, this means that *properly speaking* only God is a substance as only God can exist without assistance from any other thing. Other things are independent (and therefore substances) in the sense that, with the exception of God who must create them and keep them in existence, they depend on nothing for their existence. This means that "the term 'substance' does not apply *univocally* as they say in the Schools, to God and to other things."[71] Adopting some terminology developed by Alice Sowaal, we can draw a distinction in Cartesian thought between "primary substances" and "secondary substances." God is a primary substance, one that exists completely independently. Secondary substances are independent in the looser sense that they depend on nothing other than God.[72]

Much like Descartes, Newton also recognizes a distinction between primary and secondary substances.[73] And he views the distinction as important for a full and suitable appreciation of God's nature. Recall from just above that Newton thought a conception of bodies capable of existing apart from God's creative act was a route to atheism. Indeed, he is willing to say that there is "almost no other reason for atheism" than the view that bodies have a complete and absolute independence.[74] Understanding that God is a primary substance while created things are mere secondary substances will instill the proper theistic outlook. Like Descartes, Newton appeals to the Scholastic apparatus of univocal and equivocal predication to clarify his position. It is only the prejudice that bodies are primary substances and can exist apart from God that "explains why the same word, substance, is applied univocally in the schools to God and his creatures."[75] God is a

[70] *CSM*, I, 210. There has been much debate about this independence criterion for substancehood and whether and how it captures Descartes' full and considered views on criteria for substancehood. For discussion of other relevant passages and positions in the literature, see Schechtman 2016.

[71] *CSM*, I, 210. For helpful background and discussion on non-univocal predication in Descartes and Cartesianism, see Ariew 2022.

[72] Sowaal 2004, page 223. Sowaal continues on to develop an account of "tertiary substances" that need not concern us here. It should be emphasized that this usage of primary and secondary substance terminology differs from the distinction between primary and secondary substances in Aristotelian thought.

[73] Shein 2012 and Kochiras 2013 offer similar approaches to Newton's equivocal approach to substances. Shein is particularly helpful in situating Newton's views against certain parts of the Cartesian background. To Shein's discussion of this background I would just add that, in addition to Descartes himself, the Cartesian Lambert van Velthuysen may have been an influence on Newton's thought in this area. Newton dog-eared the page discussing "Of Being in General" in a treatise by van Velthuysen which does not mention "substance" as an equivocal term but which does discuss related issues. See *NL* #508.

[74] *DG* 47.

[75] *DG* 47.

substance in a different way, or after a different fashion, than created things. While created substances depend on God, God depends on no other substance.[76]

Eric Schliesser has rejected an approach along these lines.[77] He claims that Newton is a substance monist in the sense that God is the only substance in Newton's system.[78] His reasoning is that Newton requires something to be both (a) fully independent and (b) active in order to be a substance. Minds and bodies, while active, are not fully independent and are therefore not substances.

Schliesser's approach seems to embody precisely the univocal approach to the term "substance" that Newton warns against. His insistence on both independence and activity as necessary for substancehood also sits very uneasily with an earlier passage in *De gravitatione* which allows that activity alone is sufficient:

> For although philosophers do not define substance as an entity that can act upon things, yet everyone tacitly understands this of substances, as follows from the fact that they would readily allow extension to be substance in the manner of body if only (si modò) it were capable of motion and of sharing in the actions of body.[79]

Newton allows that we can conceive of space existing apart from all other things, but his considered view is that it requires at least one other being in order to exist. So while complete independence and activity may be Newton's criteria for qualifying as a *primary* substance, activity alone seems to be his criterion for (secondary) substancehood.[80] Schliesser's proposal also seems to run contrary to Newton's very common practice of referring to created substances as substances.[81] Indeed, there is a section of *De gravitatione* that is rather insistent on this point:

[76] There is an important point to be made here about God's dependence on space and time. John Carriero and I have both argued that Newton is (problematically) committed to God's being dependent on space (and, by analogy, on time) in order to exist and act. See Carriero 1990, pages 127–128, Connolly 2015a. Of course, Newton adamantly denies that space is a substance, so the formulation above can stand.

[77] See Schliesser 2011a. See also Jalobeanu 2007, page 110; Benítez 2012, page 116; Peterman *forthcoming*.

[78] Thus Schliesser could be seen as endorsing a form of *token* monism. Janiak, rejecting this view, has argued for a kind of *type* monism in Newton. Janiak 2008, pages 163–164. This also, on my view, conflicts with Newton's clear claim about the nonunivocity of substance. But see also Janiak's helpful discussion of conceptions of substance in Newton at Janiak 2008, pages 140–141.

[79] *DG* 36, underlining added. Note as well that the next sentence, *pace* Schliesser, seems to embed the presupposition that all people unproblematically and correctly conceive of bodies as substances: "And on the contrary, they would hardly allow that body is substance if it could not move, nor excite any sensation or perception in any mind whatsoever."

[80] There is also a passage in which Newton suggests that being the subject of and sustaining qualities is sufficient for something to qualify as a substance. See *NCS* #40.

[81] Kochiras, whose views on the issue of primary and secondary substances line up neatly with the account I give here, emphasizes this point in a critique of Schliesser's views. Additionally, she cites the first portion of the block quotation I offer below. See Kochiras 2011, pages 170–171.

> These beings will not be less real than bodies, nor (I say) are they less able to be called substances. . . . Nor will they be any less than substances, since they will likewise subsist and acquire accidents through God alone.[82]

Newton here emphasizes that something can properly be considered a substance, even while depending on God. Given all this, I think we can easily reject the idea that Newton is a substance monist.

This distinction between God and secondary substances in Newton's thought opens up a host of questions about the *nature* of the dependence relation or relations that obtain between God and created beings. This is, to my mind, among the most important topics in Newton's metaphysics. But my view is that these questions about the nature and extent of God's activity and involvement in creation would take us slightly far afield from an analysis of God's existence and substantial features. We will therefore save these questions about the dependence of secondary substances on God for the discussion of bodies in Chapter 5.

2.6. Conclusion

This chapter has examined Newton's teleological arguments for God's existence. These arguments take as their primary focus certain biological and celestial phenomena. And they have as their conclusion not only claims about God's existence but also claims about God's wisdom and providential choice in the creation and framing of the world. A somewhat cryptic remark in a letter to Bentley also opened an investigation into Newton's belief in God's absolute independence and the fact that all created beings are fundamentally dependent on God. This distinction between primary substances (which are independent of any other substances) and secondary substances (which depend on God but on no other substance) marks a major fault line in Newton's metaphysics of substance.

Newton's reflections on divine ontology were not limited to claims about God's existence. The next chapter examines his thought as it relates to divine necessity, the architecture of the divine substance, and God's relationship to space and time.

[82] *DG* 43.

3
Divine Necessity and Divine Attributes

3.1. Introduction

The last chapter focused on God's existence and independence. This chapter focuses on issues relating to God's necessity, unity, and attributes. While Newton believed that God existed of necessity, he denied that God always acted out of necessity. Newton's claims that the universe betrays evidence of free and providential divine choice set the stage for a discussion of whether and to what degree he may have been engaging with Spinoza and Spinozism. Next, Newton often affirms that there are multiple divine attributes but also that God is simple and unified. This, of course, raises familiar questions about the relationship between various divine attributes and the divine substance. Consideration of divine attributes also raises questions relating to space and time because some commentators claim that Newton understood space and time to be divine attributes. This chapter charts a path through these issues.

3.2. God as a Necessary Being

Thus far we have seen a number of Newtonian arguments to the effect that God exists. And the arguments also speak to God's overwhelming power, unsurpassed wisdom, and providential care. Indeed, God is a substance after an entirely different fashion than bodies or minds; God is a primary substance. All that being said, the arguments evinced thus far have only established that God does exist or that God's existence is *actual*. But Newton is clear that he understands God to be a *necessary* being. As the General Scholium boldly asserts: "It is agreed that the supreme God necessarily exists."[1] Or, as Newton writes in "Irenicum, or Ecclesiastical Polyty tending to Peace": "We are to beleive that there is one everliving omnipresent omnipotent omniscient invisible God the creator of heaven & earth & of all things therein. . . . That He is what he is by the necessity of nature & that all other things are what they are by the power of his will."[2] This section will consider Newton's

[1] *Principia* 942.
[2] Keynes Ms. 3, 35.

Newton's Metaphysics of Substance. Patrick J. Connolly, Oxford University Press. © Patrick J. Connolly 2025.
DOI: 10.1093/9780191959240.003.0003

approach to divine necessity on its own terms. The next will look at divine necessity in connection with the observed contingency of the created world.

We can begin by asking what Newton understood God's necessity to consist in. J.E. McGuire has claimed that "For Newton, God is a necessary being in that it is causally impossible that he be otherwise in any respect whatever."[3] This is a good starting point and seems to me to be, broadly speaking, correct. We can approach this understanding of divine necessity from, at least, two different directions. One way to make sense of God's inability to be otherwise, a way that draws on the causal language in McGuire's formulation, is to note that no other being can cause God to change in any way. As the General Scholium points out, God is able to be intimately present to other objects without being affected by them: "In him all things are contained and move, but he does not act on them nor they on him."[4] Even more revealing of this line of thought is a passage from *CPQ*: "It is true God is as far as vacuum extends, but he, being a spirit and penetrating all matter, can be no obstacle to the motion of matter; no more than if nothing were in its way."[5] Whereas the claim in the General Scholium was about the *actual* fact that God suffers nothing from the motion of bodies, *CPQ* makes the *modal* assertion that bodies cannot causally interact with God in the way they do with other bodies. This lines up with *necessary* existence in that it captures the sense in which God is not *contingent* on anything else for his existence. A tree can be destroyed by a human. It is therefore a contingent being insofar as it, in some sense, relies on the human not interfering in order to continue existing. Any secondary substance can be destroyed by God. Secondary substances are thus contingent in the sense that they rely on divine will to continue existing. This is not the case with God. God's existence is not subject to any outside force. So one way of cashing out God's necessity has to do with the way in which God's existence is not contingent.

A second way of approaching McGuire's claim has to do with God embodying a certain kind of constancy or immutability. It is not just *external* beings that are unable to change God, something internal to God's nature also demands or secures unchangingness. The Newtonian texts touching on divine necessity leave much to be desired. But it seems to me that this approach lines up more neatly with what little he does say about divine necessity. Newton, as will be discussed more below, held that what was necessary must be somehow the same in all times and at all places: "Whatsoever necessarily exists, exists always and everywhere, since the law of necessity is the same in all places and times."[6] And so he often draws a tight link between divine necessity, divine simplicity or uniformity, and existence in all places and at all times. Section 3 of *Tempus et Locus*, for example, can be seen as

[3] McGuire 1978a, page 503.
[4] *Principia* 941.
[5] *CPQ* 409.
[6] McGuire 1978b, page 123. Newton thus seems to channel what we would now call the "statistical model" of modality.

Newton's effort to articulate a version of these concepts that makes them mutually compatible.[7] And the idea seems to be that it is the necessity that determines the uniformity, ubiquity, and omnipresence. As he writes in "Of the Religion of the Iews & Christians" God is "by consequence of necessity everlasting & every where immoveably. . ."[8] A fuller quotation from the passage in the General Scholium quoted above also highlights the idea that omnipresence and uniformity depend on necessity: "It is agreed that the supreme God necessarily exists, and by the same necessity he is *always* and *everywhere*. It follows that all of him is like himself."[9] The same type of logic seems to be on display in sec. 7 of *Tempus et Locus* which emphasizes the necessity of the divine universal presence: "The most perfect idea of God is that he be one substance, simple, indivisible, alive and making live, necessarily existing everywhere and always."[10] And it also shows up in his "Twelve Articles on Religion": "The father is immovable, no place being capable of becoming emptier or fuller of him than it is by the eternal necessity of nature."[11] So, taken from this direction, God's immunity to change is not just a function of being immune to causal influence from other beings. There is also something in the divine nature that secures an unchanging being across all times and places.

The previous paragraphs have attempted to characterize Newton's understanding of divine necessity. We can now ask about Newton's reasons for believing that God *is* a necessary being. Simply put, I have been unable to find in Newton a *philosophical* argument for God's necessary existence. The closest thing I have discovered is a note that appears several pages before the beginning of *CPQ* in Newton's Trinity Notebook. Folio 83r contains notes on Descartes with references to the associated pages in his texts. One of these is a reference to "The argument drawn from necessary existence included in God's essence explained."[12] On the back of this page, 83v, there is the following:

> Axiom: It is a contradiction to say, that thing does not exist whose existence implies no contradiction, and being supposed to exist must necessarily exist. This reason is that an immediate cause and effect must be in the same time and therefore the preexistence of a thing can be no cause of its past existence (also because

[7] For some discussion of this point and background, see McGuire 1978a, page 496.

[8] "Of the Religion of the Iews & Chrsitians," 1r. The location of this manuscript is currently unknown. It was part of Lot 255 in the 1934 Sotheby's sale and part of Lot 511 in the 2004 Sotheby's sale, where it was not sold.

[9] *Principia* 942.

[10] McGuire 1978b, page 123. Translation slightly altered.

[11] Keynes Ms. 8, 1r. See also Yahuda Ms. 9.2, 139r: "For God is alike in all places. He is substantially omnipresent, & as much present in ye lowest Hell as in ye highest heaven." and Yahuda Ms. 15.3, 67r: "But the father is a pure spirit invisible intangible & immovable, being alike in all places & incapable of incorporation."

[12] *CPQ* 465.

the after time does not depend on the former time). It is only from the essence of it that a thing can perpetuate its existence without extrinsic help, which essence, being sufficient to continue it, must be sufficient to cause it, being the same reason of both.[13]

McGuire and Tamny argue that the content here derives from considerations in Descartes' *Meditations* as well as the second and fourth sets of objections and replies.[14] They are certainly right to highlight these thematic links. But I am somewhat intrigued by the possibility that Newton meant to be working toward his own articulation of an argument here, rather than merely parroting Descartes. For one thing, this note is written in ink somewhat darker than the notes on Descartes on the previous page. And unlike those notes, there are no page references to Descartes' text given here. There is also room at the bottom of the previous page, enough that if Newton were merely continuing his notes on Cartesian themes he could have started writing this text there.

Whatever we conclude about Newton's doxastic attitude with respect to this note at the time he wrote it, we can safely conclude that the argument it gives is not one he would have continued to endorse. The reason is that the success of the argument trades on its proponent having epistemic access to the essence of the necessary being. Descartes, with his innate idea of God, is in a position to run an argument of this type. He is able to clearly and distinctly perceive that God's essence contains existence as one of its necessary features. Newton, by contrast, adamantly denies that we have knowledge of God's substance or essence. So he is not able to get the argument up and running. This, in turn, may explain the fact that there are no other versions or formulations of this argument in Newton's corpus.[15]

Ultimately, Newton may have felt no argument for God's necessary existence was required. Or it may have been his view that Scripture offered considerations which pointed to God's necessity. Some passages to which Newton was particularly drawn emphasize, for example, that God fills all places and times. And Revelation 1:8, in Newton's paraphrase from sec. 6 of *Tempus et Locus*, affirms that God is "He that was and is to come."[16] This, combined with the considerations above, might have served as irrefutable evidence that God had certain features—temporal omnipresence and ubiquity—that could only be possessed by a necessary being. But this would not constitute a strictly philosophical argument as it relies on a Scriptural premise.

[13] *CPQ* 465.
[14] *CPQ* 137–139.
[15] For a somewhat different take on Newton's rejection of the ontological argument, see McGuire 1978a, especially sec. 4.
[16] McGuire 1978b, page 121.

3.3. Divine Necessity and Divine Free Will

There is one seam of Newton's thought that links up issues relating to necessity, God, and the created world in interesting ways. And it involves a contrast Newton draws between "brute" necessity and God's necessary existence, the latter of which is compatible with God acting freely. The best-known version of this Newtonian argument comes from the General Scholium: "No variation of things arises from blind metaphysical necessity, which must be the same always and everywhere. All the diversity of created things, each in its place and time, could only have arisen from the ideas and the will of a necessarily existing being."[17] Roger Cotes, editor of the second edition of the *Principia,* in which the General Scholium first appeared, included a very similar line of reasoning in his Preface. After suggesting that those who believe the universe arose out of "some necessity of nature" are guilty of sinking "to the lowest depths of degradation," he offers the following argument:

> On this supposition [that nature arose out of necessity], matter will also be uniform everywhere, for variety of forms is entirely inconsistent with necessity. . . . Surely, this world—so beautifully diversified in its forms and motions—could not have arisen except from the perfectly free will of God, who provides and governs all things.[18]

Both Newton and Cotes juxtapose two potential explanations for the origin of the universe: necessity and the free choice of God. And both point toward the diversity or variety found in nature as evidence of the inadequacy of the former explanation.

Further insight into these passages can be gained by noting that they bear a striking resemblance to arguments developed and deployed by Samuel Clarke. One of Clarke's major goals was to establish the claim that the universe existed as the result of a choice made by a free agent. This involved negative arguments against the claim that the universe could be explained by appeal to either random chance or absolute necessity. And, as Timothy Yenter has shown, one key Clarkean strategy involved an emphasis on the manifest diversity present in the world.[19]

Why, for Clarke, is absolute necessity incapable of accounting for the creation of our world? The answer lies in an incompatibility between a feature of necessity and a feature of the world. The relevant feature of necessity is precisely the one

[17] *Principia* 942. Newton wrote something quite similar at Gregory Ms. 247, 14r: "Solum enim ens intelligens vi voluntatis suæ secundum intellectuales rerum ideas propter causes finales agendo varietatem rerum introducere potuit." Transcribed at Ducheyne 2012, page 41, footnote 195. "For only an intelligent being, by means of the force of its will, acting according to the intellectual ideas of things for the sake of final causes could introduce a variety of things." See also the application of this general principle to the difference between the sun and the other planets in Newton's correspondence with Bentley. NCS #398.

[18] *Principia* 397.

[19] Yenter 2014.

highlighted by Newton, the fact that it is the same at all times and in all places. As Clarke says at one point, "But whatever is self-existent, must of necessity exist absolutely in every place alike, and be equally present everywhere."[20] The relevant feature of the world is the fact that it contains manifest diversity and change. There is supposed to be no way for something so manifestly different at different places and times to be the product of something so perfectly uniform and unchanging.[21] This appears to be the exact line of argument sketched briefly by Newton and Cotes.

Clarke's *A Demonstration of the Being and Attributes of God* was based on a set of Boyle Lectures given in 1704 and was first published in 1705. This naturally gives rise to the question of whether Newton and Cotes were influenced by Clarke in the second edition of the *Principia*. The question has been raised, in different ways, by both Timothy Yenter and Eric Schliesser.[22]

Fortunately, there are earlier texts from Newton that settle the issue. One comes from Newton's undergraduate years. While making notes on Daniel Stahl's *Axiomata philosophica* Newton wrote the following:

> The same thing, insofar as it is the same thing, always does the same thing. Insofar as it is the same, i.e. keeping itself in the same manner. This axiom extends to causes that act by nature or necessity, but not to free causes (such as God is). The same [thing] always does the same, i.e., inasmuch as in it lies, but because of diversity of objects, it would do different things.[23]

Nothing in Newton's note marks a significant departure from Stahl's text. But, at bare minimum, the seed of the argument is present. Section 8 of *Tempus et Locus*, part of which was quoted above, shows Newton continuing to develop the thought at some time in the early 1690s.[24] Indeed, it contains a full articulation of what is only gestured at in the General Scholium.

> Whatsoever necessarily exists, exists always and everywhere, since the law of necessity is the same in all places and times. And hence every diversity of things which is found at diverse places and times, was not of necessity, but drew its origin from the will of a necessarily existing being. For only an intelligent being by the power of its will acting on account of final causes could bring about the

[20] Clarke 1998, page 34.

[21] For a more detailed and nuanced presentation of this argument and the relevant texts from Clarke, see Yenter 2014.

[22] Yenter 2014, note 13 and Schliesser 2012, sec. 6.

[23] CUL Ms. Add 3996, 46v. Translation taken from *CPQ* 18 but slightly altered. For Newton's source material, see Stahl 1645, pages 94–98. On these notes, see *CPQ* 17–19 and Wallace 1988.

[24] I follow here the dating of this manuscript argued for in McGuire 1978a. Levitin 2022, pages 802–808 is confident the text was produced much later, in the 1710s. I am unpersuaded by the considerations Levitin offers. I continue to be impressed by the deep thematic links to *DG* and see the negative arguments as being geared at Descartes, rather than Leibniz.

variety of things. But variety is especially found in bodies, and the bodies which impinge on the senses are the fixed Stars, the Planets, the Comets, the Earth and their parts.[25]

Here we have the full argument. Diversity is a manifest feature of the world; we see it on earth and in the heavenly bodies. Necessity requires sameness in all places and at all times. This means that the world cannot be "of necessity" but rather was created as a result of God's free choices. This should be sufficient to show that, even if Clarke's work was what prompted Newton to revisit arguments of this form, Newton had arrived at the argument independently of Clarke.

The connection to Clarke is still intriguing, however, because it points to one other possibility. The texts of Newton's formulations of the argument, viewed on their own, make it difficult to specify the target of the argument. But the same is not the case for Clarke; it is very clear he took himself to be arguing against Spinoza. Might the same be true for Newton? More generally, how much did Newton know of Spinoza, either through reading Spinoza's own works or via some other source? And how carefully had he considered Spinoza's views?

Iliffe and Schliesser have both claimed that Spinoza loomed large in Newton's mind. The former suggests he "must have been keenly aware" of Spinoza's views, likely as early as the 1680s or 1690s.[26] The latter claims that in the General Scholium Newton "lends his authority to the project of refuting Spinoza."[27] He allows that Newton does not explicitly mention Spinoza but maintains that it would still have been clear, given the anti-Spinozist arguments authored by More and Clarke, that Newton intended to target Spinoza by echoing their claims. To my mind, the situation is much less clear. First and foremost, the Newtonian arguments in the General Scholium that Schliesser points to do have affinities with certain arguments of Clarke and More. But they do not at all seem to match up with the specific anti-Spinozist arguments centering on the possibility of motion that

[25] McGuire 1978b, page 123. McGuire 1978b, pages 126–128, notes the connections between the latter parts of *Tempus et Locus* and the General Scholium, but does not focus on this particular argument.

[26] Iliffe 2017, pages 20–21.

[27] Schliesser 2012, page 454. See also Schliesser 2013a and Schliesser 2021, ch. 8, though note that this latter shifts the focus from Spinoza to "Spinozism" and is agnostic about whether Newton was directly acquainted with Spinoza's work. Hurlbutt 1965, page 127, also claims that Newton aims to target Spinoza. And McGuire 1978a, page 493, sees it as probable that Newton knew Spinoza's views and sought to distance himself from them. A more cautious approach comes from Gorham who claims that Newton probably never read Spinoza. See Gorham 2011a, footnotes 52 and 129. Peterman *forthcoming* quite sensibly remains agnostic on the issue. Popkin 1988, page 83, Manuel 1974, pages 84–85, and Peterman *forthcoming* raise the interesting possibility that, while he was at Cambridge, Newton may have learned about Spinoza as result of his interest in Biblical criticism. They note that Isaac Barrow's library, a library Newton had access to and cataloged, contained a copy of Spinoza's *Tractatus Theologico-Politicus*. But they do not provide any positive evidence that Newton was acquainted with the book. Levitin 2022, page 670, is deeply skeptical of any links between Spinoza's and Newton's thought.

Schliesser reconstructs earlier in his paper.[28] Certainly, if Newton intended to echo those arguments he might have signaled this more clearly.

As for Clarke, Newton did own the 1706 edition of Clarke's *Demonstration* and did know Clarke personally. So we might expect that by the time of the General Scholium he had given some thought to Spinoza's views and Clarke's criticisms of them. But, as we have just seen, Newton had formulated the relevant arguments long before Clarke's Boyle Lectures, the publication of the *Demonstration*, and, in all likelihood, before he met Clarke.[29] So, I think we ought to be hesitant about the claim that Newton was merely endorsing or amplifying Clarkean critiques of Spinoza.

With More the situation is more complex. Certainly, More was a major influence on Newton and Newton read a great deal of More's work. But, in this connection, the key text seems to be More's *Confutatio* (1687). While Newton owned many of More's works, the *Confutatio* is absent from his library. There is, of course, no doubt it would have been readily available to him in Cambridge and that he might well have read it or encountered its arguments indirectly. But without more by way of solid evidence that Newton knew these Morean arguments, I again think we should be hesitant to claim that he intended to channel them in the General Scholium.

Two further considerations also counsel caution. First, Cotes' version of the argument in the Preface seems to take *Descartes*, rather than Spinoza, as its target.[30] Immediately after insisting that it is the divine will, rather than necessity, that governs all things Cotes goes on the attack:

> From this source [the perfectly free will of God], then, have all the laws that are called laws of nature come, in which many traces of the highest wisdom and counsel certainly appear, but no traces of necessity. Accordingly we should not seek these laws by using untrustworthy conjectures. . . . He who is confident that

[28] Though it should be noted that a portion of Cotes' text elided in the quotation above *does* contain one version of this argument.

[29] I have been unable to determine with precision when Newton and Clarke first met. But Clarke did not matriculate at Cambridge until 1691. Admittedly, his tutor at Caius, John Ellis, was a personal acquaintance of Newton's. And admittedly, Clarke did show early promise in natural philosophy. E.g., he defended a proposition from Newton in a disputation for his bachelor's degree in 1695. But it is still hard to imagine Clarke coming to Newton's attention as an undergraduate, i.e., before the composition of *Tempus et Locus* (at least as dated by McGuire). It seems more likely that Newton first learned of Clarke after the publication of the latter's translation of Rohault in 1697.

[30] There is some correspondence in which Newton, Bentley, and Cotes discuss this issue. See NCS #980, #984, and #985. Leibniz is initially mentioned as another potential target, though at first seemingly only with reference to the priority dispute, and only later with respect to more substantive issues. But Bentley and Newton asked that Leibniz not be mentioned by name. Cotes tells Newton that "De Cartes & others" are his target. None of this definitively rules out that Cotes meant to be targeting Spinoza. But it is striking, that in all this discussion on the runup to Cotes' preface, Spinoza is never named by any of the key protagonists. See Westfall 1980, page 749 and Iliffe 2007, page 122, for the view that Leibniz is the main target of Cotes' preface.

> he can truly find the principles of physics, and the laws of things, by relying only on the force of his mind and the internal light of his reason should maintain either that the world has existed from necessity and follows the said laws from the same necessity, or that although the order of nature was constituted by the will of God, nevertheless a creature as small and insignificant as he is has a clear understanding of the way things should be.[31]

The suggestion here that the laws of nature might be understood to follow with necessity from something necessary fits cleanly with Descartes' deduction of the laws of nature from divine immutability (a necessary feature of a necessary being) in *Principles,* Part II, secs. 37–42.[32] And the suggestion that humans could deduce these a priori using the force of the mind and the internal light of reason also seems like a clear reference to Cartesian methodology. More generally, the argument here targets those who would build *systems of physics* on claims about necessity and reason rather than on solid empirical grounds. And Descartes, unlike Spinoza, was recognized as being the originator of such a system. Finally, we can note that the first horn of the dilemma presented at the end of the quotation does present a serious challenge to Descartes but not to Spinoza, who happily endorses just that position. Given that Cotes saw the argument as a refutation of Descartes, we might acknowledge that it is at least possible that Newton intended to do the same and had Descartes, rather than Spinoza, in mind in the General Scholium.

Second, and somewhat more generally, we can note that nothing about this argument forces us to conclude that Newton had Spinoza in mind while deploying it. The reason is that the argument has rich antecedents in the history of philosophy that do not take Spinoza as their target. Maimonides' *Guide of the Perplexed,* for example, offers a structurally similar argument. The context is his consideration of the venerable question of whether the universe has existed eternally or whether it was created in time. His stated goal is to establish the latter position.[33] One stage of his argument involves arguments against "necessity" which would support an eternal universe. Obviously Maimonides does not have Spinoza as one of his targets. Instead, he is arguing against a certain kind of Aristotelianism, one which holds that "that which exists has proceeded from the Creator in virtue of a necessity; that He, may He be exalted, is a cause and this world an effect and it was necessary that this should be so."[34] And the question which leads off his challenge to this view is precisely one about the manifest diversity of the world: "What then is

[31] *Principia* 397–398.

[32] *CSM*, I, 240–243.

[33] Of course, there is a longstanding question in Maimonides interpretation regarding the sincerity of the "stated" views in the *Guide* and the role of esotericism in his work. See Seeskin 2017, sec. 7. For current purposes, we can set these difficult issues to one side and merely pursue the argument offered at face value. But for a discussion of these themes in connection with Maimonides' arguments concerning creation, see Weiss 2010.

[34] Maimonides 1963, volume 2, pages 302–303.

the cause of the differences between the individuals of every species?"[35] If the hypothesis of necessity were true, Maimonides claims, then we could expect a high level of regularity or homogeneity in the world. As it happens, however, we see a high level of diversity and, indeed, arbitrariness in the world. This diversity and arbitrariness can only be attributable to the free choice of an agent. One particularly striking feature of Maimonides' argument, given our present concerns, has to do with the fact that he singles out certain celestial phenomena as evidence of arbitrariness. Specifically, he points to the direction of the planets' orbits, the speed of the planets, and the placement of the fixed stars. These, he claims, could only be the product of divine choice.

Here we find an argument that neatly parallels Newton's. It appeals to just the type of phenomena Newton's does and it does so in service of just the same conclusion Newton hopes to establish. And it self-consciously targets a hypothesis that the universe is the product of necessity. As is the case with More's *Confutatio*, Newton owned several works by Maimonides but not the *Guide of the Perplexed*.[36] This makes it very hard to claim that Newton knew of and was influenced by Maimonides' views on this point, thought we cannot rule this out.[37] The important point is just that Maimonides shows that the argument had, by the time Newton formulated it, a philosophical life quite independent of Spinoza and Spinozistic considerations. And so it is entirely possible that Newton did not have Spinoza in mind when he deployed the argument.

To be clear, none of this is intended as an outright denial of the claim that Newton took Spinozism to be his target. To my mind this is a tantalizing possibility. And I think the questions of when and how Newton learned of Spinoza's work, how carefully he engaged with it, and what his reaction to it might have been are good ones and well worth pursuing. My worry is just that, on the basis of the evidence currently available, they are not ones we can provide secure answers to. So I think the appropriate attitude at present is an agnostic one.

There is one further text which bears on this issue and deserves discussion. It comes from a section of *Of Natures obvious laws & processes in vegetation* headed "Of God."[38] This text is normally considered to have been written between 1670 and 1675.[39] This is important for questions regarding Spinoza, as the *Ethics* was written in roughly the same period, but not published until 1677 in his *Opera Posthuma*. Here is the relevant selection:

[35] Maimonides 1963, volume 2, page 304.
[36] See *NL* ##1018–1022.
[37] For discussion of Newton's engagement with Maimonidean thought, see Popkin 1992.
[38] The precise relationship of "Of God" to the larger text it is embedded in presents some puzzles. For opposing views, see Newman 2019, pages 170–173 and Snobelen 2021, pages 176–177.
[39] For discussion of the dating, see Newman 2019, page 139, note 6.

> Of God. what ever I can conceive without a contradiction, either is or may bee made by something that is . . . Arg 2. The world might have been otherwise then it is (because there may be worlds otherwise framed then this) Twas therefore noe necessary but a voluntary & free determination that it should bee thus. And such a voluntary [cause must bee a God] determination implys a God.[40] If it bee said the world could bee noe otherwise then tis because tis determined by an eternall series of causes, thats to pervert not to answer the 1st proposition: For I meane not that the [Earth] might have been otherwise notwithstanding the precedent series of causes, but that the whole series of causes might from eterity [sic] have beene otherwise here, because they may bee otherwise in other places[41]

The text clearly shows Newton working through issues of necessity and divine choice in creation, albeit in a less developed way than he was seen doing above.

Iliffe has voiced serious concerns about the validity of Newton's argumentation here. On his reconstruction, "Newton concluded that the current state of this world, resulting from God's 'voluntary & free determination y^{t} it should be thus' was evidence of both the latter's existence and benevolence" but this result was achieved "with not a little circularity."[42] I disagree. First, I see the passage as only arguing to the conclusion that God exists and being silent as to divine benevolence. Second, I see the claim about voluntary and free determination as part of the conclusion, rather than as a premise. If I am right then the argument will be rescued from circularity.

Here, then, is my reconstruction. Newton starts with modal considerations. What is necessary is what must be the case, or what could not have been otherwise. Whatever is conceivable without a contradiction is possible. It is conceivable that things could have been different than they are. I can imagine a world in which the coffee cup on my desk is two inches to the left of where it currently sits, I can imagine the Earth having two moons, and so forth. Therefore, it is possible that things could have been different. If it is possible that things could have been different, then things are not necessary. Up to this point, the argument is on relatively solid footing.[43] What happens next is slightly less convincing. Newton presupposes there are only two possibilities for the origin of the world: necessity or free choice. This, conveniently for Newton, elides the possibility of random chance. With necessity ruled out, Newton can conclude that the universe has the form it

[40] Newton here appears to have first written "And such a voluntary cause must be a God." He then added the square brackets, partly covering the period, and continued on to add "determination implys a God."

[41] Dibner Ms. 1031 B, 4v.

[42] Iliffe 2017, page 97. The reconstruction offered at Dobbs 1991, pages 113–114, may be subject to the sort of circularity that concerns Iliffe.

[43] One could, of course, deny that conceivability and possibility line up in the way Newton suggests, or deny that we actually can clearly conceive of things being different. My point is just that Newton is appealing to reasonable premises held by many of his contemporaries and the argument is logically valid.

does as a result of free choice. And he then infers that this free choice entails the existence of a God. This last argumentative step is highly compressed, but without too much work the necessary suppressed premises could be brought out and made to justify it.[44]

This argument does not trade specifically on manifest diversity in the world. It takes, in some sense, a more direct route to its conclusion. But it does still try to target the idea that necessity requires a certain kind of uniformity; necessary causes must, according to Newton, always and everywhere operate in the same way. And if the universe were governed by necessity there would be no possibility for the differences we clearly know to be possible. So, I think we can see this as an ancestor, or at least a relative, of the arguments discussed above. The argument shows that Newton was eager to resist a kind of necessitarianism and establish divine choice as the cause of the world. And, barring some revision to the dating of the manuscript or new evidence that Newton may have learned about Spinoza's views before they were published, it predates any exposure he could have had to Spinoza's views.

This still leaves open, of course, that some versions of the argument constructed above were written with Spinoza in mind or were designed to target Spinoza's views. Again, my view is that questions regarding Newton's awareness of and attitude toward Spinoza are good ones, well worth investigating. At the same time, I think we should not be too hasty to conclude that Newton had Spinoza as his target given the long history of anti-necessitarian arguments both in philosophy, generally, and in Newton's own thought.

3.4. Deflationary and Non-Deflationary Interpretations of Divine Substance

We have seen that Newton believed that there were arguments attesting to God's existence. And, related to some of those arguments, he believed that God was the primary and only truly independent substance. Moving on from questions about God's existence and status relative to other substances, we should now consider the ontology of the divine substance. What, metaphysically speaking, *is* God? On the one hand, given some of the methodological constraints discussed at the outset of Chapter 2, Newton's answer is likely to be somewhat limited. As he said, we have no idea of the substance of God. On the other hand, Newton does make a number of revealing remarks related to the ontology of God that are worth exploring.

[44] Newman reads this section of the text as part of Newton's ongoing engagement with and rejection of Descartes' arguments for God's existence, and sees links between this note and those on necessity and the ontological argument in *CPQ*. See Newman 2019, pages 171–173. As my reconstruction may indicate, I am not convinced that Newton means to be engaging Cartesian thought here, and am unpersuaded that this text does have clear overlap with *CPQ*.

God is, if anything, certainly a substance. And Newton is very often happy to talk about divine attributes. Relatedly, he also believes God has various powers. So, a correct understanding of Newton's divine ontology will involve a correct understanding of the relationship between these various things.

One of the most developed accounts of substance, and in particular the divine substance, in Newton's thought comes from Hylarie Kochiras.[45] She offers a deflationary account of Newtonian substances according to which substances just are their characteristic attributes. This means that "the divine substance consists in omnipotence or creative power" or, in a different formulation, "the divine substance just is the active power to create."[46] Kochiras' proposal is rich and insightful. And in the course of developing her interpretation she does well to push back against some commentators who have suggested that Newton is an eliminativist about substance.[47] Talk of substance is so commonplace in Newton's writings that it is hard to imagine that he hoped or intended to eradicate the concept. Further, *De gravitatione* offers criteria for substancehood and deploys some technical apparatus familiar from traditional philosophical discussions of substance. For example, as discussed above, and as noted by Kochiras, Newton seems eager to invoke the familiar distinction between primary and secondary substances. But while I am in full agreement with Kochiras' critique of substance-eliminativist readings of Newton, I believe we should be hesitant about her positive claim that God is nothing over and above an attribute or that God just is God's creative power.

While I am highly sympathetic to Kochiras' claims that created substances are nothing over and above collections of powers, I do not believe the same can be said of the divine substance. Put differently, my view is that Newton believed that God had not just attributes but also some essence or substantial nature undergirding those attributes.

One initial reason for thinking this comes from Newton's distinction between primary and secondary substances. In the discussion above, the distinction was framed in terms of *dependence*. God, the primary substance, is truly independent. Minds and bodies, created secondary substances, are dependent on God for their existence. But there is another strand of thought in Newton's distinction. It is less pronounced in the text, but nonetheless important for our current concerns. Recall that Newton believed atheists, and those in danger of falling into atheism, affirm things of created substances that they ought to only affirm of God. Specifically, they claim substances have "a complete, absolute, and independent reality in themselves."[48] Here independence is only part of what is at issue. Similarly, a few

[45] Kochiras 2013.

[46] Kochiras 2013, pages 328 and 345. On page 342 she, pointing to Newton's claims about the unknowability of divine substance discussed above, adds the very reasonable caveat that "this is a concept of the substance insofar as we can know it."

[47] She targets, in particular, Stein 2002 and Slowik 2009.

[48] *DG* 47.

sentences below, when offering a more accurate characterization of created substance, Newton writes that "whether you consider [a] its degree of dependence or [b] its degree of reality" it falls somewhere between God and accidents.[49] This indicates that dependence is not the only way of differentiating between the substantial nature of God and creatures. God also has a greater degree of reality. My view is that God has a greater degree of reality, in part, because unlike secondary substances God is more than a collection of attributes or powers.

Newton does take issue with the claim that secondary substances have some substance in which their properties or attributes inhere. One reason is that he says it will endanger any substantial difference between mind and body. If minds are constructed by adding mental attributes to "substance" and bodies are constructed by adding bodily attributes to "substance" then it would seem that minds and bodies are really the same thing—"substance"—differently modified. As Newton writes:

> [I]f bare substances do not have an essential difference, the same substantial forms or attributes can fit both, and render them by turns, if not at one and the same time, mind and body. And so if we do not understand that difference of substance deprived of attributes, we cannot knowingly assert that mind and body differ substantially.[50]

Newton does believe, at least at the time of *De gravitatione*, that mind and body differ substantially and so rejects the idea that they are "bare substances" with properties or attributes added on. So far so good. But Newton then goes on to initiate a contrast between bodies and God on this front.

Suppose we persist in the error of thinking bodies are bare substances with bodily attributes inhering in them. A further problem with this, Newton says, is that this will leave those like us attributing "no less reality in concept (though less in words) to this corporeal substance regarded as being without qualities and forms, than they do to the substance of God, abstracted from his attributes."[51] One way to read this claim is as Newton *affirming* that God does indeed have some bare

[49] *DG* 47.

[50] *DG* 46. Intriguingly, Henry More, who is often seen as a *positive* influence on Newton's views, may be one of the targets of this Newtonian argument. See More 1653, page 11 and More 1659, pages 115–120, especially 112. For helpful discussion, see Duncan 2022, pages 115–125. With respect to the form of the argument, Locke deploys structurally similar considerations in the *Essay*. He says that if we can apply the concept of substance univocally to God, finite spirits, and to bodies then "will it not thence follow, That God, Spirits, and Body, agreeing in the same common nature of *Substance*, differ not any otherwise than in a bare different modification of that *Substance*; as a Tree and a Pebble, being in the same sense Body, and agreeing in the common nature of Body, differ only in a bare modification of that common matter; which will be a very harsh Doctrine." Locke 1975, page 174. Isaac Watts also uses a form of this argument. Perhaps surprisingly, given the text just quoted, Locke is Watts' target. See Watts 1733, pages 49-50.

[51] *DG* 46.

substance in addition to some attributes. This is the fundamental difference between God and bodies, and the reason God has more reality "in concept" than do bodies. Put differently, Newton here affirms that *even abstracted from the divine attributes* God's substance has more reality than bodies. And this indicates that God is more than just attributes; God has some substance. The mistake lies in affirming that bodies, setting aside their qualities and forms, might have some further being or substance in which those qualities and forms inhere. And the content of the mistake is affirming that bodies have a substance in the way that God does.

Recall from above that others emphasize parts of the text surrounding this passage in which Newton claims there is something nonsensical or contradictory about claiming that bodies could consist of properties inhering in an underlying substance. They see the problem as lying in the very idea of a substance. We are now in a position to locate the contradiction elsewhere.

A closer look at Newton's argumentation here will show that he only intended for this argument to apply to created substances. What Newton hopes to establish is that an understanding of created substances on which they consist of both a substratum and attributes inhering in that substratum will leave no suitable way to make sense of their dependence on God. The problem is not in conceiving of a substratum. Newton seems to think that this is so easy that, in fact, "almost all of us, through negligence, are accustomed to have in our minds from childhood" this precise view of bodies.[52] The problem is in conceiving of this substrate as being both created and dependent. This is what Newton finds nonsensical or contradictory.[53]

The rest of the paragraph clarifies why this is so. Newton here reverts to a traditional conception of substances and accidents. Substances are things capable of existing independently. Accidents are not capable of existing independently. Accidents, therefore, rely on their substances to "sustain" them.[54] And God sustains created beings no less than, on the traditional scheme, substances were believed to sustain accidents. Thus, we ought not think of created substances in the traditional way. But they are nonetheless properly called substances. And for this reason, "substantial reality is to be ascribed to these kinds of attributes [of created things] which . . . do not need to be inherent in a subject, rather than to the subject which we cannot conceive as dependent." None of this amounts to an argument against the very idea of a substance acting as an inherence base for attributes. Rather, it amounts to an argument against a *dependent* being consisting of a substance with attributes inhering in it. So there is nothing problematic in the

[52] *DG* 47.

[53] Newton's reasoning here has some resonances with the arguments found at Digby 1644, pages 39–40. Newton owned the 1658 printing of this work: *NL* #516.

[54] This same approach is on display in a text in which Newton criticizes transubstantiation. One of the many "miracles" required for each act of consecration is "To make the accidents subsist by themselves, [that is become substances]." Yahuda Ms. 14, 20r. Bracketed content in original.

idea that God has some substance in which God's attributes inhere, even if created beings do not.

One key text that can motivate this interpretation comes from *Tempus et Locus*. In sec. 6 of that text, Newton offers a list of contrasts between what he takes to be the mysterious, and therefore dangerous, conceptions of God affirmed by theologians and what he takes to be the correct understanding of God. Relevant for current purposes is one of his claims about omnipotence. It is agreed on all hands that Newton rejects virtual divine presence, the view that God is present only insofar as God can exercise divine power, in favor of actual or substantial divine presence.[55] But the phrasing of this point in *Tempus et Locus* is telling. He asks whether we should believe with the theologians "that God is everywhere as regards power and nowhere [as regards] substance" or whether we should, more correctly, believe "that God's power should subsist everywhere in the divine substance [as its] proper sub[strate], and exist [nowhere separately], and have [no medium] by which it be propagated from its proper substance [into external places]."[56] The passage is obviously complex. But the most straightforward reading has Newton differentiating between divine power and the divine substance and has the divine substance as a substrate in which the divine power inheres. At a minimum, it seems hard to square Newton's general approach and framing of the question here with a belief that the divine substance just is divine power.

There are three textual points from *De gravitatione* to clarify before moving on. First, Newton warns that people who conceive of both God and created substances in the traditional way, when abstracting the attributes away from both, will assume they are equal in the sense of being bare substances. They will "conceive of both, when considered simply, in the same way; or rather they do not conceive of them, but confound them in some common apprehension of an unintelligible reality."[57] Second, Newton says that the attributes of created substances do not need to be inherent in a substance and we should not ascribe them to some substance we cannot conceive of as being dependent "much less form any idea of it."[58] One might worry that, whatever the cogency of the argumentation in the previous paragraphs, these passages suggest Newton harbored doubts about the intelligibility of the very idea of substance.[59]

My view, however, is that if Newton holds that God was the only substance that consists of something beyond a collection of attributes or properties then these are precisely the claims we could expect him to make. If God has an essence underlying

[55] He makes the point explicitly in the General Scholium, see *Principia* 941. He also affirms it in his Treatise on Revelation, see Yahuda Ms. 9.2, 139r.
[56] McGuire 1978b, page 121.
[57] *DG* 46.
[58] *DG* 47.
[59] McGuire 1978a, page 472, e.g., appeals to the first of these passages to claim that Newton "denies the coherence of the notion of substance itself." Gorham 2011b, page 27, also claims that Newton rejects the intelligibility of substances as having substrata.

the other divine features, then it would not be surprising for it to be precisely the sort of thing that humans are unable to conceive. This should be clear for two reasons. First, God is infinite and human minds are finite. This should guarantee that the divine essence outstrips our ability to understand it. Second, God's essence is not the sort of thing we can come into contact with and, given the broadly empiricist contours of Newton's epistemology, it is therefore not the kind of thing we could expect to form clear ideas of. *Pace* the nativist Descartes who felt God had left an impress on our mind just as craftsperson stamps her work, Newton claims there "is no direct sense and there are no indirect reflected actions by which we know innermost substances; much less do we have an idea of the substance of God."[60] I read this as Newton claiming that there is a thing, the substance of God, which we have no idea of. It may be unintelligible to us, but it is nonetheless a reality.

The third textual point is more difficult and more directly relevant to the question of whether or not God has a substance. The relevant sentences are as follows:

> In the same way, if we should have an idea of that attribute or power by which God, through the action of his will alone, can create beings, we should readily conceive of that attribute as subsisting by itself without any substantial subject and [thus as] involving the rest of his attributes. But while we cannot form an idea of this attribute, nor even of our proper power by which we move our bodies, it would be rash to say what may be the substantial basis of mind.[61]

> *Ad eundem modum si Ideam Attributi sive potestatis istius haberemus quo Deus solâ voluntatis actione potest entia creare: forte conciperemus Attributum istud tanquam per se sine aliqua subjectâ substantiâ subsistens et involvens cætera ejus attributa. Sed interea dum non hujus tantùm Attributi, sed et potestatis propriæ quâ nostra corpora movemus, non possumus Ideam efformare: temeritatis esset dicere quodnam sit mentium substantiale fundamentum.*

It should be clear why Kochiras appeals to this passage to support her reading. She takes it to say that a clear conception of God's creative power would not involve the conception of that power inhering in a substance. Howard Stein offers a similar analysis claiming that the passage shows Newton's "rejection of the notion of 'substance' as having reference to what he calls an 'unintelligible' support or subject of attributes."[62] And a similar view can be found in the work of McGuire, Dempsey, Slowik, and DiSalle.[63]

[60] *Principia* 942.
[61] *DG* 47.
[62] Stein 2002, page 281. See also Stein 1970, page 276 and Stein 2023, page 134.
[63] McGuire 1982, pages 145–146; Dempsey 2006, page 432; Slowik 2009, page 451; Slowik 2013, pages 443–444; DiSalle 2016, page 49.

I think, however, that both textual and contextual considerations align this passage much more closely with my own understanding of the divine nature. Textually, it is worth emphasizing that an alternative translation of Newton's Latin is available, and most likely required, here. First, one could try to read "forte" in the sense of strength as implying that "we should readily" conceive of the attribute as described. But "forte" could also be read in the sense of chance. To read "forte" with the sense of strength, it would have to be understood as an adjective modifying "Attributum"; this, however, would give something like "we should conceive of that *powerful* attribute" rather than "we should *readily* conceive of that attribute." Instead, I think "forte" ought to be read in the sense of chance, implying that we "perhaps" or "may" conceive of the attribute as described. This would fit with it either being an adverb, we "perchance" or "perhaps" could do this. Or it would fit with the ablative of means, that it is by means of luck or chance that we would be able to do this. Second, the translation above either elides the "tanquam" in the sentence or reads it as an intensifier to say that we could conceive of the attribute in "just" this way. But I believe the "tanquam" is better read as introducing a note of hesitancy and indicating that we could conceive of the attribute "as it were" in this way.[64]

So an alternate reading might have Newton saying something more like the following. If we could formulate a clear conception of God's creative power then perhaps it would be possible to conceive of this power as existing—so to speak—on its own, apart from the divine substance. But given that we cannot do this, we probably should not speculate too much on the substantial nature of our minds. This reading, far from showing that God's substance just is God's creative power, makes it possible that God's creative power is just one of the many attributes that inhere in the divine substance.

Contextually, we can ask what is meant to be happening in the passage. What is Newton argumentatively in search of here? If we ask this question, then the last clause becomes very important. Newton is trying to articulate some point about the substantial nature of minds. This is something obscured by Kochiras' reading. Indeed, it is hard to see what role a point about human minds could play in her interpretation of the passage.

Zooming out, however, we realize that Newton is trying to teach us something about *secondary* substances. He is working to help us throw off the childish prejudice that they consist of attributes inhering in substances. The way to do that is to ascribe substantial reality to the attributes themselves. As an intuition pump, he asks us to imagine space, which we already can think of as existing independently, and he points out that *if* we could think of space as having activity (as having, per Descartes' suggestion, mobile parts) then we would have no hesitation in affirming it was a substance. Given our verdict in that case, we should not have any trouble

[64] Böhme's German translation nicely captures the points made here with its inclusion of "vielleicht." Newton 1988, page 69. Biarnais's French does the same with its inclusion of "peut-être." Newton 1985, page 60.

affirming that spatially located powers of motion are a substance. Newton carries on and introduces the target passage above with "in the same way" suggesting he will offer an argument with parallel structure. And we can read him as doing precisely this.

If we could imagine God's creative attribute existing alone then we would not have any trouble thinking of it as having some substantial reality. So, by analogy, if we could conceive clearly of that "power by which we move our own bodies" we might ascribe substantial reality to it and allow that it constituted our mind. We might be satisfied to say that our mind was a substance even if it consisted of a power or collection of powers rather than a power or set of powers inhering in a substance. Of course, we do not have clear ideas on this front so should refrain from speculation.

Note, however, that if the arguments are really parallel, then there is an important upshot for what Newton says about the conception of God's attribute as existing independently. The intuition pump in the body case was something confused. If we *could* clearly conceive of the parts of space being mobile, then we would say they were corporeal substances. But, *pace* Descartes, we cannot. Similarly, if we *could* clearly conceive of this divine attribute existing apart from the divine substance and other attributes, then we might conceive of such a thing as being a mental substance. But, in the same way we cannot imagine parts of space as themselves mobile, we might not be able to form a clear conception of God's creative power divorced from the divine substance and other attributes.

To put this contextual point differently, Newton's goal in this passage is to offer us resources for thinking about secondary substances. He is helping us understand why we should still affirm that secondary substances have substantial reality and deserve to be called substances even though they lack an independently existing substratum. But given the context, what is important is that he is teaching us about *secondary* substances. Kochiras reads him as making a point about substance in general, as making claims meant to apply to both primary and secondary substances. But there is no need to read him in this way. Indeed, to do so runs the risk of adopting precisely the univocal approach to substance that Newton warns us against.

3.5. God as Substance and Attributes

I have just argued that there is more to God's substantiality than the divine attributes. I now want to contest the specifics of Kochiras' claim that God just is God's creative power. My worry is that, on the one hand, this suggestion is too deflationary and, on the other hand, it is not deflationary enough. First, I worry that the suggestion is too deflationary insofar as creative power is not the only divine attribute, there are many others. Surely these are also part of God's substance. At the same time, there is a worry about not being deflationist enough. If the claim is that God is nothing over and above the divine attributes, in the way my laundry pile is nothing over and above various articles of clothing, then God will be composite or will display a certain kind of complexity. But Newton was committed to the doctrine of divine simplicity. For example, in *Tempus et Locus* he writes that "The most perfect idea of God is that he be one substance, simple, indivisible."[65] So it seems that in addition to claiming God is reducible to divine attributes we also require an account of the way those attributes are related to one another or reduce to just one thing. Kochiras seems sensitive to both of these concerns and writes the following: "Although Newton does not mention other divine attributes, such as omniscience, his view presumably includes the belief that God is identical to that attribute, which is in turn identical to omnipotence and creative power."[66] While I think Kochiras is right to try to address these issues, I find this claim puzzling.

Even if the other attributes are not specified, Newton is clearly committed to their existence. The passage that is the target of Kochiras' exposition, on both her interpretation and mine, entails that thinking about God's power to create involves thinking about "the rest of his attributes." And certainly in later texts, such as the General Scholium, Newton is happy to mention omniscience alongside omnipotence and refer to the "thinking substance of God."[67] These other divine attributes seem to be core features of Newton's conception of God.

So Kochiras is correct to acknowledge that there are further divine attributes. But doing so seems to mark an abrupt departure from her central contention that God *just is* creative power. It seems to concede that God is more than or includes something beyond creative power. It is one thing to claim that God *is* or is *identical* to God's creative power. This would just be to claim that God is not really distinct from creative power. And this would leave open the possibility that God is identical to some further things. But Kochiras' claim that God *just is* God's creative power (or that God is constituted by God's creative power) is a much stronger claim. That is the claim that God is nothing over and above creative power. And this would seem to rule out God also being identical to some further attributes. Put differently, Kochiras' allowance that there are additional divine attributes makes it hard to understand the sense in which God just is creative power. At a minimum, if there are further attributes and they are identical to creative power, then it should be equally permissible to claim that God just is or is constituted by any one of those attributes.

[65] McGuire 1978b, page 123. The General Scholium also displays a commitment to divine simplicity. Newton claims that it follows from God's being necessary that "all of him is like himself." *Principia* 942. Section 3 of *Tempus et Locus* can also be read as displaying Newton's commitment to divine simplicity. See McGuire 1978a, page 496, for an analysis along these lines.

[66] Kochiras 2013, page 343.

[67] *Principia* 941. For other ascriptions of omniscience which put it on a par with omnipotence and omnipresence, see Yahuda Ms. 7, 1v; Yahuda Ms. 15.5, 98r; Keynes Ms. 3, 35.

Starting with Kochiras' more general interpretive claim that substances are to be identified with their characteristic attributes, one might take her to be claiming that creative power is God's *primary* or *characteristic* attribute in the sense that other attributes reduce to it. But while God's creative power is surely an important feature of God and a key *differentia* between God and creatures, it is hard to see what would justify the claim that it is uniquely important. And it is not obvious how other divine attributes are fully reducible to or comprehensible in terms of creative power. God's omniscience, immutability, and moral authority, for example, are not obviously understandable with sole reference to creative power. And this, in turn, makes it difficult to see how they could be identical to that power.

Notably, Kochiras is not the only commentator who faces difficulties as a result of Newton's commitment to multiple attributes and also to divine simplicity. It also sits uneasily with J.E. McGuire's claim that Newton is opposed to views on which attributes are identical with God's nature and thereby identical with the divine essence and existence. As McGuire notes, as a consequence of rejecting these identifications "Newton must answer the objection that his conception of predication makes God divisible and a composite nature of substance and attribute."[68] McGuire's general argumentative strategy is to emphasize the way in which God is immutable to change. The analysis offered is rich and plausible. But it is hard to see how it addresses the central charge that a robust distinction between different attributes and between attributes and substance is incompatible with a commitment to divine simplicity. After all, immutability and simplicity are distinct concepts. And while it might be possible to derive immutability from simplicity it is less clear how to derive simplicity from immutability. The problem McGuire points to is a serious one.

My view is that McGuire is right to insist on Newton's commitment to divine simplicity but wrong to claim that Newton rejected an identity between attributes and the divine substance. On closer examination, McGuire's argument for this second claim is underwhelming. The argument is motivated by two passages. The first comes from the General Scholium:

> He is not eternity and infinity, but eternal and infinite; he is not duration and space, but endures and is present. He endures forever, and is everywhere present; and by existing always and everywhere, he constitutes duration and space. Since every part of space is *always*, and every indivisible moment of time is *everywhere*, certainly the Maker and Lord of all things cannot be *never and nowhere*.[69]

[68] McGuire 1978a, page 479. See also page 483: "on his theory of predication, Divine attributes are not identical, a conception essential to the view that God is a simple and unitary being without distinction . . . . So Newton has to find grounds for Divine simplicity other than the claim that in God being and unity are one."

[69] *Principia* 941.

The second comes from drafts of an "Avertissment au lecteur" that Newton prepared for the 1720 Des Maizeaux printing of the Leibniz-Clarke correspondence:

> when the Hebrews called God MAKOM place, the place in wch we live & move & have our being & yet did not mean that space is God in a literal sense. For they used to speak of God by figures & allusions & put space for his omnipresence by a figure. And so space & duration are by the writers of Logicks & Metaphysicks called quantities & with respect to God's ubiquity & omnipresence they have a nearer relation to the Predicaments of *Ubi* & *Quando* then to that of Quality & therefore where they are called Qualities the Reader is to understand it with a figure.[70]

McGuire takes these passages to show that ontic distinctions exist between the different divine attributes and between the divine attributes and the divine substance. Here is the relevant analysis:

> These passages give a clear expression of Newton's general conception of predication. As he denies that God *is* Eternity and *is* Infinity, so he denies that God is identical with Wisdom and Power. In other words, he opposes those theologians who hold that God's attributes are strictly one and the same with Divine essence and existence, as well as identical among themselves.[71]

Simply put, I find it hard to read these passages as supporting McGuire's conclusion.

As is so often the case with Newtonian texts, the fact that these passages concern space and time makes them difficult. But this very point is also worth emphasizing. The passages concern precisely that: space and time. They also concern God and God's relationship to space and time. But it is hard to read them as being about *divine attributes in general.* McGuire claims that what is true of eternity and infinity will be true of wisdom and power. But nothing in the passages touches on omniscience or omnipotence or gives any indication that those attributes might be at stake in the discussion. So already McGuire's effort to draw general conclusions about divine attributes from the passages seems questionable.

Again, the passages are very specifically focused on space and time and God's relationship to space and time. As McGuire himself convincingly argues, and as will be discussed below, space and time are *not* divine attributes for Newton.[72] Given

70 Koyré and Cohen 1962, page 101. Koyré and Cohen offer a helpful discussion of the context and content of these drafts.

71 McGuire 1978a, page 479. The passage continues: "According to Newton, Divine nature is just, wise, eternal and infinite. These characteristics are attributable to God in a perfect manner (in comparison with those which are appropriately ascribable to finite things), but none is identical with his nature."

72 See McGuire 1978a, sec. 2.

this, it is unsurprising to see Newton rejecting an identification of God with space and time in both passages. In the first, Newton's claim is straightforward: *God is not duration and space*. In the second there is also a straightforward claim (even the ancients did not claim space was God in a literal sense) but also a more complicated one: We must be very careful to avoid thinking that space and time are qualities of God in any normal sense. So far so good. But all that this can support is the claim that Newton rejected an identification of God with space and time. It tells us nothing about the relationship between divine attributes and the divine substance.

With respect to space and time the relevant divine attributes can be called, taking a cue from the second passage, omnipresence and ubiquity. God is not space and time but God is present in and fills all spaces and all times. My diagnosis of McGuire's proposal is that he has conflated, on the one hand, the divine attributes of omnipresence and ubiquity with, on the other, space and time. This means that when Newton denies an identity between God and space and time, McGuire reads him as denying an identity between God and the attributes of omnipresence and ubiquity. But, particularly in the second passage, it is really the relationship between the attributes and space and time that is at issue. The ancients were speaking non-literally because they referred to space when they should have referred to the divine attribute of omnipresence; they "put space for his omnipresence." And the philosophers call space and time *qualities of God*, but they actually stand in a different relation *to the divine attributes of omnipresence and ubiquity*, so they are not literal qualities of God. Space and time are just the "whereness" or at what place (*ubi*) and "whenness" or at what time (*quando*) of the divine attributes of omnipresence and ubiquity. Read in this light, the passages seem largely silent on the relationship between God and the attributes.

So I see McGuire's proposal for robust ontic distinctions between the divine attributes and the divine substance as undermotivated. And given Newton's concern for divine simplicity, it seems safer to attribute to him the view that he saw the divine attributes as being identical to one another and as being identical to the divine substance.

Where does this leave us with respect to Newton's views on the ontology of God? On the basis of the discussion above, the following story seems correct. Newton believed God consisted of a substance and numerous attributes. The divine substance is unknown to us and is, indeed, unknowable by us. Additionally, God is a simple substance. All of this is attested to either by Newton's texts or the arguments of this chapter. What is less clear, on this picture, is how to understand the relationship between the substance and the attributes in a way that preserves divine simplicity.

One option for Newton on this front, an option that would have him aligned with some mainstream contours of early modern theology, would be to claim that the attributes are grounded in, or flow from, the divine essence in such a way that they are only rationally distinct from the essence. This would entail that they are only rationally distinct from one another. This is a tidy picture.[73] And, as we will see in Chapter 6, Newton did have some facility with early modern theories of distinctions. But my suspicion is that this was not his considered position on divine ontology. As discussed above, he may have seen attempts at reconciling his various commitments using a tool like the rational distinction as a dangerous imposition of philosophy onto theology.

A second option, one which I suspect is much closer to Newton's true views, has it that the relationship between the substance and the attributes and their mutual identity is something that is opaque to human understanding. Given Newton's well-founded view that the substance of God was unknowable by us, it would perhaps not be very surprising if its relationship to the divine attributes and its simplicity also surpassed our understanding.[74] Newton may have been channeling Henry More's thought on this point who offered it as an Axiom that:

> There are some Properties, Powers and Operations, immediately appertaining to a thing, of which no reasons can be given, nor ought to be demanded, nor the Way or Manner of the cohæsion of the Attribute with the Subject can by any meanes be fancyed or imagined.[75]

This axiom is supported by the thought that if the subject or substratum of a thing is itself unconceivable then surely the link between it and the substance's properties or attributes will also be inaccessible to the mind. As for divine simplicity and the identity of the attributes with one another and with the divine substance, Newton does gesture at a kind of mysterian approach in the General Scholium where he claims that: "all of him is like himself: he is all eye, all ear, all brain, all arm, all force of sensing, of understanding, and of acting, but in a way not at all human, in a way not at all corporeal, in a way utterly unknown to us."[76] And it gets a slightly

[73] Janiak 2008, page 154, canvases something along these lines but stops short of endorsing it. Gorham 2011b develops a conception of Newtonian divine attributes along, roughly, these lines. But Gorham does so with the purpose of arguing that space and time are divine attributes. As will be discussed below, however, Newton rejects the view of space and time as divine attributes. Because Gorham's article is tightly focused on space and time, it does not offer an analysis of Newtonian divine attributes like omniscience and omnipotence. So, at least to my mind, it remains unclear whether such an interpretation could find adequate support in the texts and be successfully defended.

[74] Insofar as this is the correct view, I am sympathetic to the claim at McGuire 1978b, page 503 that "Unfortunately, neither Newton nor Spinoza explain a serious difficulty: what is meant by saying that God's properties and attributes are 'in' his substantial nature? Simply to show that God's properties are casually inseparable from his nature, gives no account of how they are connected 'in' God. But without this account, the relevant kind of causation cannot be understood."

[75] More 1659, page 11. Axioms 1 and 2, on page 6, and Axiom 8, on page 10, are also relevant for this issue.

[76] *Principia* 942. This passage bears a close resemblance to two roughly contemporaneous passages from Newton's Drafts on the History of the Church. These texts are more complex in that it is less clear Newton is speaking in his own voice, but the overlap with the General Scholium is significant and the commitment to divine simplicity more straightforward. The first reads: "And all these things are but one thing in several degrees & have place only in the mind of man. They err therefore in ascribing to God the affections & passions of men & making him a compound. For God is not as man, nor are his

more developed treatment in a manuscript expositing 2 Kings 17:15–16 from the 1670s: "To celebrate God for his eternity immensity omnisciency & omnipotency is indeed very pious & the duty of every creature to do it according to capacity, but yet this part of Gods glory, as it – – – – almost transcends the comprehension of man so it springs not from the freedom of Gods will but from the necessity of his nature."[77] As a *rational* ontology of God, this position has obvious drawbacks. But, as Newton says, "the best notions any of us have of him are imperfect enough."[78]

3.6. God, Space, and Time: Attributes

One further question about the ontology of God deserves consideration.[79] Several authors in the secondary literature have suggested that space and time are—in some sense—part of the divine being. Most commonly, commentators claim that space and time are divine attributes. As seen in the previous section, Newton's claims about the relations between God, the divine attributes of ubiquity and omnipresence, and space and time can be confusing and require careful treatment. That being said, Newton's clear position was that space and time are not divine attributes. They have a different standing in his ontology. This section will make the case that they are not divine attributes. The next one will work to unpack Newton's

thoughts like ours. He is simple & not compound. He is all like & equal to himself, all sense all spirit, all perception all Ennœa, all λόγος all ear, all eye, all light. He is all sense which cannot be separated from it self, nor is there any thing in him which can be emitted from any thing else." Yahuda Ms. 15.6, 109v. And the second reads: "They err therefore saith Irenæus in ascribing to God the affections & passions of man & making him a compound. ffor God is not as man nor are his thoughts like ours. He is simple & not compound. He is all like & equal to himself all sense all spirit all perception all Ennœa all λόγος all ear, all eye all light. He is all sense which cannot be separated from it self, nor is there any thing in him which can be emitted from any thing else." Yahuda Ms. 15.7, 128r. These, in turn, bear a close relation to a passage from Clement of Alexandria copied by Newton earlier, in the 1670s, and reading in part: "non divisus, non dissectus, non migrans e loco in locum; sed ubique semper existens, nec ullo modo circumscriptus: totus mens, totus lumen paternum, totus oculus, omnia videns, omnia audiens, omnia sciens, potestate potestates perscrutans." *Notes from Petavius on the Nicene Council*, Keynes Ms. 4, 14r. Westfall 1980, page 318, translates this as: "not divided, not disjointed, not moving from place to place, nor in any way circumscribed but existing everywhere always; all mind, all paternal light, all eye, seeing everything, hearing everything, knowing everything, examining powers by his power." On the theme of textual resonances in the General Scholium, see also De Smet and Verelst 2001.

[77] Yahuda Ms. 21, 2r. Note as well the tightness of the connection between God and the divine attributes in a passage on the next folio: "As for his attributes he knows those can scarce be given to any other, because so soon as men do ascribe any of his attributes to any thing they worship, they doe thereby mean the supreme God." These texts are related to a sermon, or series of two sermons, Newton may have been required by statute to preach as a master of arts, possibly in Trinity College chapel. See Westfall 1980, page 354 and Levitin and Mandelbrote 2019, pages 341–342. Though, see Feingold 2021, page 183, for an alternative understanding. Levitin and Mandelbrote offer further analysis of the notes as does Iliffe 2017, page 140.

[78] Yahuda Ms. 21, 3r.

[79] Much of the material in this section is drawn from Connolly 2015a, secs. 2 and 3.

claim that they are common affections of all beings just as beings. I will focus my discussion on space. But similar claims should hold, mutatis mutandis, for time.[80]

The first point to establish is that Newton believed space to be a really existing being.[81] This serves to distinguish him from many of his contemporaries. And the most important text for this is, unsurprisingly, *De gravitatione.* There Newton makes explicit his belief that space is a really existing entity. He writes that space "has its own manner of existing which is proper to it and which fits neither substances nor accidents."[82] Newton has two reasons for denying that space is a substance. The first is that space "is not absolute in itself."[83] Newton's suggestion here is that space is an affection of being; this means that in order for space to exist some other being must also exist. Newton's second, and more important, reason for denying that space is a substance is the fact that space does not act. Indeed, space is causally inert, incapable of acting on its own or of being acted upon.[84] Newton's argument that space is not an accident is as follows. There is no one particular thing which is necessary for the existence of space. We can imagine a variety of objects coming into and going out of existence without any change to space's nature. This shows that space does not inhere in and does not depend on any one particular object for its existence and so cannot properly be considered an accident.[85]

So Newton's claim is that space does exist as a real being, but that it is a very special kind of being. Put differently, I read Newton as arguing that the set of existing beings is not exhausted by the subsets of substances and accidents. There are really existing beings which can properly be classified as neither. This is why Newton speaks of space as "having its own manner of existing which is proper to it and which fits neither substances nor accidents."[86] Space exists in a *way* or *manner* which is different from the way or manner in which substances and their accidents do. This is why it can be classified as a different type of being.[87]

[80] This follows the common practice of treating space and time as rough equivalents. For some discussion of this practice in the early modern period, see Palmerino 2011 and Gorham 2012. For a treatment of time in early modern British philosophy, as it were, on its own merits, see Thomas 2018.

[81] "Being" is, of course, a difficult term. A primary goal of this section is to make more precise the sense in which Newton understood space to be a being. In the relevant texts, Newton offers us four ontological categories: substance, being, accident, and affection. Newton is very clear that space is neither substance nor attribute. He also is clear that it is an affection. But, membership in these categories is not mutually exclusive. God, e.g., is a substance as well as a being. I will argue that space is a being as well as an affection.

[82] *DG* 35.

[83] *DG* 36.

[84] Zvi Biener has provided a careful and powerful argument to the effect that Newton ultimately abandoned this commitment. See Biener 2017. For reasons that will be discussed in Chapter 6, I am ultimately unpersuaded by this argument. But it merits serious consideration.

[85] See the related discussion in Janiak 2008, page 145.

[86] *DG* 35.

[87] It follows from this that I agree with others in the literature who have rejected a substantivalist-relationist dichotomy as inappropriate for classifying Newton's views. See, e.g., Janiak 2008, page 155; Slowik 2009; Slowik 2013.

Consider another passage from *De gravitatione* in which Newton makes it explicit that he believes space is a really existing entity: "it is usually believed that these spaces [the ones devoid of body] are nothing; yet indeed they are spaces. Although space may be empty of body, nevertheless it is not in itself a void; and something is there, because spaces are there, although nothing more than that."[88] Here Newton takes himself to be correcting a common error. Some people consider empty space to be devoid of being. But this is incorrect, because space itself is a being, in empty space there is some *thing*, namely space itself. Note as well the relative independence that space is given in this quote. Space is not presented as having derivative being. The explanation of the being present in empty spaces stops at the mere existence of space: "spaces are there, although nothing more than that."

At one other point in *De gravitatione* Newton affirms the real existence of space. The occasion for this is a comparison of space to Aristotelian prime matter. Newton writes that "They differ, however, in that extension (since it [involves] "what" and "how constituted" and "how much") has more reality than prime matter."[89] Here Newton has a Scholastic concept of prime matter in mind: something which is either pure potency or has the lowest possible level of actuality. Newton very carefully points out that space has more reality than prime matter. So again, we get the view that space is a really existing entity, one which has its own (relatively high level of) reality and its own manner of existing.

Newton's view that absolute space was a really existing being also surfaces in the *Principia*. The Scholium on space and time makes the following claim: "Absolute space, of its own nature without reference to anything external, always remains homogeneous and immovable."[90] This passage is important for two reasons. The first is that Newton suggests here that space has its own nature. The language of natures offers us some prima facie reason for thinking that Newton thinks space is a being.[91] Second, Newton's suggestion that we can consider space in isolation from other beings ("without reference to anything external") suggests it has some being. If space were just a product of relations between material objects or a pure potentiality or nothing, then considering it apart from other beings would be impossible.

Perhaps the most decisive piece of evidence for the claim that Newton believed that space was a real being is a passage from sec. 3 of *Tempus et Locus*. This manuscript closely resembles certain passages from *De gravitatione* and suggests that

[88] *DG* 41.

[89] *DG* 44. This passage is helpfully contrasted with the following from a draft of the Queries: "[P]rime matter which is neither a thing, nor possessed of quality, nor a thing which can be measured, is not a phenomenon." CUL Ms. Add 3965, f. 422r. Translation taken from McGuire 1966, page 220.

[90] *Principia* 408.

[91] On this point see McGuire 2007. He argues that Newton adopted some relevant components of Descartes' theory of natures. Specifically, Newton adopted the view that natures, in and of themselves, determine the structure or content of some being. So, in claiming that space either is or has a nature Newton is claiming that space is a conceptually independent and positive being.

Newton's position on space remained consistent from the time he composed *De gravitatione* through the first edition of the *Principia*. In this manuscript, Newton explicitly claims that space is a being (*ens*): "For it is a single being, [most simple], and most perfect [in its] kind."[92] So, several important Newtonian texts give us reason to think that he understood space to be a really existing being.

Before moving on, I want to discuss the work of one commentator who might be seen as offering a dissenting view. Andrew Janiak has denied that Newton believed that space was a being:

> [I]n addition to being contingent, space and time are also not beings, and are therefore not necessary beings. Unlike contingent beings such as ordinary objects and human minds, which . . . are substances that bear characteristic actions, space and time are causally inert and therefore fail a crucial Newtonian criterion of substancehood. Space and time are mere affections, and not genuine substances. Thus for Newton, space and time are merely contingent and are mere affections.[93]

It will be worthwhile to spend a few moments analyzing this claim. When properly understood, I believe that this quotation, far from giving us reason to doubt that space is a being, can actually help us to understand the exact sense in which Newton believed space to be a being.

Janiak is perfectly correct to note that for Newton space and time are not substances. His characterization of space as an affection is also unproblematic. And Janiak is perfectly correct to think that we can infer that "*x* is not a substance" from the claim that "*x* is a mere affection." What is troubling about this passage, however, is that Janiak also seems to think that we can move from the claim that "*x* is not a substance" to the conclusion that "*x* is not a being." This is the move which I think is questionable. Janiak offers us no independent reason for thinking that Newton restricts the set of beings to the set of substances.

My position is that when we consider the polemical goals of *De gravitatione* we realize that Newton was consciously trying to avoid just this sort of restriction. As Janiak himself points out, a major goal of *De gravitatione* is to develop a metaphysic which is non-Cartesian (and non-Scholastic). In the course of doing this, one thing which Newton does is reject the claim that all entities can be classified as substances or attributes. Indeed, Newton identifies a special metaphysical category (affections) for entities (space and time) which are neither substances nor attributes. Given all of this, and given Janiak's careful work to show the sense in which Newtonian space is not a Newtonian substance, I think we can reformulate Janiak's claim that space is not an entity. We can more charitably read Janiak as claiming that space is not a *substance*. If we read him this way then we can both agree that

[92] McGuire 1978b, page 117.
[93] Janiak 2008, page 147.

Newton denies that space is a substance, but does think that it is an entity (in the looser sense of the term).

Here is another way into the same point. "Being" used as a count noun is, quite obviously, a vague term. Janiak is operating with a restricted or narrow usage of the word. His move from the claim that space and time are not substances, while still allowing that they are affections, to the claim that they are not beings indicates that he makes a certain degree of independence essential to being a being. By contrast, I am adopting a more permissive or lightweight conception of being. On my view, a being is just anything that exists. Consider Scholastic accidents. On most views (and bracketing out the miraculous case of the consecrated host), Scholastic accidents cannot exist without their substances. We encounter red walls and red tomatoes and red shirts, but we do not and cannot encounter red or mere redness existing on its own. For Janiak, it seems, this would entail that the accident of redness is not a being. On my view, Scholastic redness is partially constitutive of the fabric of the universe. Unlike, say, Cartesian redness it is not fully reducible to (the texture of) a material object. Like other Scholastic accidents, it is an entity that inheres in a substance. I am therefore happy to call it a being. If all of this is correct, then Janiak and I have a merely verbal dispute.

Understanding that, for Newton, space was an entity distinct from God is, in a sense, difficult. It is difficult because it requires understanding that Newton differed from some of his contemporaries who held very similar views. Specifically, it involves understanding how Newton was different from his most immediate predecessor on the topic, Henry More, and how he was different from his most immediate successor on the topic, Samuel Clarke. These writers identified God with space. I will show that this is a position which Newton consciously and consistently rejected.

Henry More's views on the nature of space were both complicated and subtle. Furthermore, they underwent significant changes throughout his career.[94] Nevertheless, I think it is uncontroversial that More's mature position held that absolute space was to be understood as a divine attribute.[95] Samuel Clarke followed More in claiming that space was a divine attribute.[96] In his 5th Letter to Leibniz he writes as follows:

> God does not exist in space and in time; but his existence causes space and time. And when, according to the analogy of vulgar speech, we say that he exists in all space and in all time; the words mean only that he is omnipresent and eternal,

[94] See Reid 2007.
[95] See, e.g., More 1995, volume 1, pages 56–58.
[96] And Clarke was joined by other early eighteenth-century Newtonians. See, e.g., Cheyne 1705, pages 19–20 and Colliber 1718, pages 218–219.

> that is, that boundless space and time are necessary consequences of his existence; and not, that Space and Time are beings distinct from him, and in which he exists.[97]

So here Clarke explicitly denies that God and space are distinct beings.[98] Given Newton's general intellectual debt to More, his personal association with More and Clarke, Clarke's general fidelity to Newton's own views, and some indirect evidence which suggests that Newton may have had a hand in Clarke's letters,[99] we might have reason to wonder if Newton also thought space was part of God.

Some interpreters have been convinced by these considerations and have argued that Newton, like Clarke and More, did identify God with space. The American Samuel Johnson, in his correspondence with Berkeley, was among the earliest interpreters to take this position.[100] John Tull Baker, in his seminal work on early modern conceptions of space and time in England, argued that Newton conceived of space and time as divine attributes.[101] In this, he was followed by Edward Grant and Edward Slowik.[102] Edward Khamara, in summarizing the Newtonian position on space, writes that "in relation to God, however, space and time are not substances but properties: and constitute respectively the divine attributes of immensity (or omnipresence) and eternity (i.e. sempiternity)."[103] More recently, Geoffrey Gorham has offered a sustained defense of the divine attribute reading and a related view can be found in Eric Schliesser's work.[104]

As indicated at the outset of this section, I want to contest these interpretations. My position is that there is decisive evidence that Newton disagreed with Clarke and More on this point and envisioned more ontological distance between God and space than exists between a substance and its attributes. I will first consider some passages which directly show Newton claiming we should resist the urge to identify space and God. Then I will show that some of Newton's other claims about space are incompatible with a reading of space as a divine attribute.

[97] Leibniz and Clarke, page 71.

[98] Part of Clarke's motivation may have been that he thought it was impossible for there to be two necessary beings. Clarke had argued for this position in Proposition VII of his *A Demonstration of the Being and Attributes of God*. See Clarke 1998, pages 35–38.

[99] For this alleged evidence, see Koyré and Cohen 1962. As noted in the Introduction, I, in agreement with McMullin 1978, page 4, note 7, am inclined to be skeptical that we have any serious evidence of Newton's role in the construction of Clarke's letters.

[100] "External space and duration therefore I take to be those properties or attributes in God.... This I take to be Sir Isaac Newton's meaning." Berkeley 1998, page 178.

[101] Baker 1930, page 3: "Both the *Principia* and *Optics* consider absolute space and time God's attributes like More and Barrow." See also pages 29–30. It is worth noting that Baker only had access to the *Principia* and the *Optics* and not to the wealth of Newton manuscripts available to scholars writing today.

[102] Grant 1981, pages 243–244 and Slowik 2009, page 451.

[103] Khamara 2006, page 25.

[104] Gorham 2011a and Schliesser 2013b. Schliesser 2013c provides a parallel argument regarding time.

One text in which Newton warns us against identifying God with space comes in the General Scholium: "[God] is not eternity and infinity, but eternal and infinite; he is not duration and space, but he endures and is present."[105] As discussed above, I read this passage as a straightforward denial that space and God are in any way the same thing ("he is not . . . space"). While it is not appropriate to say of God that God *is* space, it is appropriate to say of God that God is *present*. Put differently, Newton in this passage wants to emphasize that God is intimately related to space, to emphasize that God is in all parts of space, to emphasize that space determines God with regard to location, but not to suggest that space is part of God. A second, similar, passage occurs in sec. 4 of *Tempus et Locus*: "By reason of its eternity and infinity space will neither be God nor wise nor powerful nor alive."[106] Newton here claims that although space and God share some features (they are both eternal and infinite), it would be a mistake to identify them or to associate space with God's power and wisdom. As Newton writes slightly later: "[N]ot everything eternal and infinite will be God."[107] So, in this passage, we get an explicit comparison between the way in which space is extended (which does not make it perfect) and the way in which God is extended (which does make God perfect).[108]

The most decisive evidence that Newton believed God and space were distinct entities was also discussed above. It comes from the draft of an *Avertissement au Lecteur* Newton wrote for an edition of the Leibniz-Clarke correspondence edited by Des Maizeaux. Here is the text from one of these drafts:

> The Reader is desired to observe that wherever in the following papers, through unavoidable narrowness of language, infinite space or Immensity & endless Duration or eternity, are spoken of as Qualities or Properties of the Substance wch is Immense or Eternal, the terms Quality & Property are not to be taken in that sense wherein they are vulgarly, by the writers of *Logick & Metaphysicks* applied to *matter*; but in such a sense as only implies them to be CONSEQUENTS of the Existence of a Being which is really necessarily & substantially Omnipresent & Eternal.[109]

Here again we have Newton insisting that the relationship between space and God is extremely tight, yet still denying that they are the same thing. Space is more like an effect of God than like a feature or part of God.

[105] *Principia* 941.

[106] McGuire 1978b, page 119. Compare with *DG* 39: "But I see what Descartes feared, namely that if he should consider space infinite, it would perhaps become God because of the perfection of infinity. But by no means, for infinity is not perfection except when it is attributed to perfect things."

[107] McGuire 1978b, page 121.

[108] On this Newtonian argument, see McGuire 1983.

[109] Koyré and Cohen 1962, page 96.

So much for Newton's direct claims about the non-identity of space and God. I now want to consider some passages which motivate indirectly against the divine attribute reading. But in order to see how these passages do this, it is necessary to say a little bit about how early modern philosophers understood attributes. There are, broadly speaking, two ways of thinking about attributes: either as (1) ways of conceiving a substance or as (2) features of a substance which inhere in it. Newton makes claims about space which are inconsistent with space being an attribute according to either conception.

The first view about attributes holds that an attribute is just a manner in which a substance is conceived.[110] Omnipotence, for example, on this view is just one way of thinking about the perfectly simple substance that is God. So space, on this view is just one way of considering the divine essence; thinking about space is really just thinking about God. Henry More seemed to favor this understanding of space as a divine attribute. He often claims that space is an "image" or a "representation" of God.[111] He means that space is one of the ways in which human minds imagine, represent, or conceive of God. I think that Newton could not have shared this view with More. The reason is that Newton believes we can conceive of space *without* conceiving of God. Consider the following passage:

> If we say with Descartes that extension is body, do we not manifestly offer a path to atheism, both because extension is not created but has existed eternally, and because we have an idea of it without any relation to God, and so in some circumstances it would be possible for us to conceive of extension while supposing God not to exist?[112]
>
> *Si cùm Cartesio dicamus extensionem esse corpus, an non Atheismæ viam manifestè sternimus, tum quòd extensio non est creatura sed ab æterno fuit, tum quod Ideam ejus sine aliquà ad Deum relatione habemus absolutam, adeòqȝ possumus ut existentem interea concipere dum Deum non esse fingimus.*

First, there are two important points to make about the standard translation, which is reproduced above. The presence of "absolutam" in "quod Ideam ejus sine aliquà ad Deum relatione habemus absolutam" seems to be entirely ignored. What Newton is saying is that we can have a fully developed or complete idea of space, even apart from God. Second, the "in some circumstances" is an unwarranted interpretative addition to the text. There is simply nothing corresponding to it in

[110] This understanding of attributes is the one employed by Gorham. For an exposition of this understanding of attributes in a Cartesian context, see Nolan 1997.

[111] E.g., "I have always been prone to think this subtile Extension . . . to be a more obscure shadow or adumbration, or to be a more general and confused apprehension of the *Divine Amplitude*." More 1668, page 106. It is worth noting that Newton never, to my knowledge, uses this sort of language to characterize space.

[112] *DG* 45–46.

the Latin. In this sense, the passage is much starker than the translation makes it seem.[113] This is especially notable when we pair this passage—which affirms that we can have a full or complete idea of space apart from God and can conceive of space even while positing that God does not exist—with a claim made later in the same paragraph: "certainly whatever cannot exist independently of God cannot be truly understood independently of the idea of God."[114] Newton's claim that we can have an absolute idea of space apart from God, in combination with this claim, shows that there is some ontological distance between them. On the view I am proposing, on which God and space are ontologically distinct, we can make perfect sense of all of this. We can conceive of space without God *because they are distinct things.*[115]

Gorham has denied that this is a decisive text against the attribute interpretation. His argument is that Newton is operating within a Cartesian context where it is possible to conceive of extension without God. Newton, according to Gorham, must reject this possibility on pain of being complicit in this atheism.[116] I think Gorham has this backwards. Descartes makes it abundantly clear that extension cannot properly be conceived apart from God, this is why extension is not, properly speaking, a substance.[117] Therefore, if the argument is to have any force at all, given that it is not Descartes' position that God and extension can be conceived distinctly, Newton himself must have believed that we can conceive of space apart from God. Further evidence that Newton is speaking in his own voice (and not channeling Descartes) comes from the fact that he claims that extension is not created and has existed eternally. Again, this is a position which is clearly Newton's and is clearly not Descartes'.[118] How, given that he thinks space can be conceived apart from God, does Newton avoid the charge of atheism? Newton claims that although conceiving of space does not require conceiving of God, conceiving of *bodies* does (as Gorham goes on to note). So what Newton is attacking in this passage is the Cartesian equation of bodies with space; it is this conflation, and not the ontological distinction between God and space, that leads to atheism.

[113] Böhme's German translation captures the more forceful sense: "weil wir ihren Begriff absolut, ohne jede Beziehung auf Gott haben, so daß wir sie uns als existierend denken können, während wir fingieren, daß Gott nicht sei." Newton 1988, page 63. So does Biarnais' French: "que parce que nous en avons une idée absolue, sans aucune relation à Dieu et qu'ainsi nous pouvons concevoir qu'elle existe tout en imaginant que Dieu n'existe pas?" Newton 1985, page 56. The Spanish translation prepared by Benítez and Robles is slightly more equivocal, perhaps as a result of their comparative work with the English (see page 23), but importantly it does contain "idea absoluta." See Benítez and Robles 2006, page 56.

[114] *DG* 47.

[115] That said, conceiving of space may require conceiving of *some* object, the important point is that that object does not need to be God.

[116] Gorham 2011a, pages 287–288.

[117] *CSM*, I, 210 and *CSM*, II, 49.

[118] See, for instance, *CSM*, I, 240. Relevant here is Newton's claim at *DG* 48 that "extension is eternal, infinite, uncreated . . . whereas body is opposite in every respect."

A second view about attributes treats them more like properties. On this view, attributes are things that inhere in their substances. On this understanding of attributes, space inheres in God. I think Newton cannot be understood as viewing space as a divine attribute in this sense. Consider this passage from *De gravitatione*: "And much less may [space] be said to be nothing, since it is something more than an accident, and approaches more nearly to the nature of a substance."[119] If Newton really did think of space as this sort of divine attribute, then presumably it would be more natural to speak of space as an accident than to speak of it as a substance. On the view which identifies God with space, space is dependent on God for its being which makes it entirely unlike a substance. But here we have Newton claiming that it is more like a substance (which has its own being) than like an accident (which inheres in a substance).

This passage is even more striking when we consider what comes just before it. Newton has delineated the two reasons why space does not qualify as a substance. He thinks it is (1) incapable of existing alone[120] and (2) incapable of action. In the analysis of these two reasons Newton provides us with an interesting counterfactual: "yet everyone tacitly understands [that substances are capable of action], as follows from the fact that they would readily allow extension to be substance in the manner of body if only it were capable of motion and of sharing in the actions of body."[121] Here Newton writes that *if* space could act then it would be substance. But this counterfactual would be nonsensical if Newton understood space as an attribute. The reason is that it is impossible for an attribute-like thing to gain the power of acting. Because attributes are ontologically linked to substances, and because substances are the proper *loci* of actions, it is nonsensical to claim that an attribute (as opposed to its substance) could act. Consider a pole-vaulter and one of her attributes, her strength. Now, it makes perfect sense to say that her strength *allowed* the pole-vaulter to jump over the bar or to say that a pole-vaulter jumps over a bar *in virtue of* her strength. But it is a serious logical error or a category mistake to claim that a pole-vaulter's strength jumps, or could jump, over the bar. And there is something bizarre about considering a counterfactual in which the pole-vaulter's strength gained the power to act while still being the pole-vaulter's strength.[122]

[119] *DG* 36. Compare with *DG* 47 where Newton says that because we can think of space without any bodies in it "some substantial reality fits this."

[120] That said, Newton does not think that space cannot exist on its own in the way that material bodies cannot exist on their own. He thinks that material bodies are dependent on God for their existence in a way that space is not dependent on God for its existence. As argued in subsequent chapters, Newtonian bodies depend on acts of divine volition for their initial creation and acts of divine conservation for their continued existence. As argued in the next section, space is an affection of all being just as being and so exists whenever any other entity exists. See *DG* 47.

[121] *DG* 36.

[122] I think that this argument against the second (inherence) way of understanding space as a divine attribute is also successful as an argument against the first (conceiving) way of understanding space as a divine attribute.

On the view I am proposing, however, we can make perfect sense of this passage. I am arguing that space is a really existing being which is neither a substance, nor an attribute, nor an accident. Space has its "own manner of existing." Now if we imagine space as a really existing being, there is no logical obstacle to imagining it as existing with causal powers. On the understanding of space as an attribute, however, for it to be considered a substance it would have to do two things. First, it would have to acquire causal powers. Second, it would have to, somehow, ontologically detach itself from its substance (God). On my view, which has space already existing as its own being, only the first of these steps would be necessary. This better fits with the sense of Newton's counterfactual claim.

The arguments presented thus far have been broadly *textual* in nature. We can also note a *contextual* consideration mentioned by McGuire.[123] Specifically, while Newton was clearly indebted to More, he also was deeply influenced by Gassendist thought (primarily via Walter Charleton).[124] More was deeply wedded to a traditional substance-accident or substance-attribute ontology. But Newton's refusal to "define extension as substance, or accident, or else nothing at all" shows that he does not think this Morean ontology is adequate to the nature of space.[125] This refusal stacks up neatly with the approach to space in the Gassendist-Charletonian tradition which gives space a certain level of autonomous being and resists any identification of space with God.

My hope is that at this stage it should be clear that Newton rejected any position which takes space (or time) to be a divine attribute.

3.7. God, Space, and Time: Affections

The arguments of the previous section were primarily negative; they argued that Newton did not see space and time as divine attributes. But if space and time are not attributes then what are they? As it happens, Newton tells us explicitly: they are common affections of all beings just as beings. And, of course, it follows from this a fortiori that they are divine affections.[126] The goal then, must be to understand what affections are as well as how and why space and time might be understood as affections.

Affection is not a familiar ontological category in the way that substance, attribute, property, and mode are. In philosophical texts from Newton's time, it most

[123] See McGuire 1978a, sec. 2.

[124] Thus, McGuire 1978a, page 464: "The basic conclusion to emerge is this: Newton is largely a Gassendist on the problems of space, time and existence."

[125] *DG* 35.

[126] My thinking on this point is indebted to McGuire 1978a, sec. 2. And I am also indebted to much of the excellent discussion in Janiak 2008, ch. 5. But I aim to expand on and add to the considerations presented in those authors and to bring more of the relevant historical context to bear on the issue.

often had a meaning akin to "passion" or "emotion" and would be used to discuss phenomena such as joy or grief. But it could also have a more general sense referring not just to minds but to all entities. In this sense, an affection is something characterizing or modifying a substance; something *affecting* it such that it is a certain way. At first pass, this might make affections sound rather like members of other ontological categories, for example, properties. For our purposes, it will be important to note the difference between attributes, on the one hand, and affections, on the other. And it will be important to focus our attention not on affections in general, but particularly on *common* affections.

My proposal is that thinkers in Newton's milieu would have been able to recognize two distinct, albeit closely related, differences between attributes, on the one hand, and common affections, on the other. First, both attributes and affections are closely related to their substances, but the relation between attributes and substances is much closer or more integral than the relation between affections and their substances. Second, attributes are *proprietary* in a way that common affections are not. An attribute is something proper, that is, it belongs only to a specific substance, where "specific" is understood to mean either a particular individual substance or to mean a substance of a certain kind (i.e. one that falls under a given species). I will first suggest that elements of this twofold distinction between attributes and common affections were present in some sources that were either read by Newton or easily available to him. I will then show that the distinction tracks Newton's language and argumentation, especially in *De gravitatione.*

One place to begin is with Descartes' *Principles.* In Part 1, sec. 48, Descartes makes a general distinction between things and affections of things. But, as he moves forward, he begins to use the language of attributes to talk about affections that are both inextricably linked to the nature of the substance and proprietary, in the sense discussed above. Section 53, for instance, notes that each kind of substance has a principal attribute that "constitutes its nature and essence."[127] And sec. 54 argues that our ideas of bodies and minds can be clear and distinct "provided we are careful to distinguish all the attributes of thought from the attributes of extension."[128] Newton clearly understood this Cartesian sense of attribute. As discussed in Section 3.4 above, in *De gravitatione* he criticizes the idea of a substratum by arguing that it would allow for the possibility of one attribute belonging to the substratum at one time and another at another time thereby making a single substance at one time a mind and at another time a body.[129] For this argument to work, Newton must be conceiving of attributes as proprietary.

We can contrast all of this with a distinction Descartes makes at the outset of his discussion of affections. Back in sec. 48, he had differentiated between things

[127] *CSM*, I, 210. See also page 215.
[128] *CSM*, I, 211.
[129] *DG* 46.

like substance, duration, order, and number "which extend to all classes of things" and the affections that are proper to specific types of beings, as perception and volition are to minds.[130] Here, we can see something like the idea of a common affection, something that affects all things universally. Descartes, of course, would deny that space could be a common affection. For him, space can only be extension and therefore must only relate to body. More generally, Descartes makes many claims in these sections and is not always entirely rigorous in his use of terms like attribute and affection. The point is just that some of the basic resources for the distinction between common affections (universal to beings of all types) and attributes (linked to specific substances and inextricable from their distinct natures) that I am working to develop would have been available to Newton from his reading of the *Principles.*

Johannes Magirus, Walter Charleton, and Isaac Barrow are three other authors known to Newton who use the language of affection. In Magirus' *Physiologiæ Peripateticæ*, Newton found a distinction between internal and external affections.[131] Internal affections—Magirus lists motion and boundaries—belong to individual bodies.[132] External affections, including place and time, are those which circumscribe and contain physical objects.[133] This, of course, would open up the possibility that external affections could be common to many different objects. As McGuire and Tamny put the point: "an external affection is something that any number of things have in common (e.g. things can endure through the same stretch of time and can exist together at a place)."[134] Charleton had claimed that "Quantity or Extension is the proper and inseparable affection of Matter, and therefore every thing hath so much of Extension, as it hath of Matter."[135] On the one hand, it can sound like Charleton is invoking what in Magirus' terms would be an internal affection. But note the final clause. The wording here would be compatible with extension being an external or common affection, some one thing that each body relates to. Barrow's second mathematical lecture of 1664 had claimed that magnitude was a common affection (*affectio communis*) of all physical beings.[136] He claimed it was deeply or thoroughly tied to the natures of these beings but did not claim that it—like a Cartesian attribute—constituted the nature of them. Instead, he claimed, it acted more like a substrate. As with the sections of Descartes' *Principles* discussed above, this material is deeply suggestive, even if it is not precisely aligned with the views I am attributing to Newton.

130 *CSM*, I, 208.
131 See the discussion of Newton's notes on Magirus and this specific point at *CPQ* 16.
132 Magirus 1642, page 26.
133 Magirus 1642, page 44.
134 *CPQ* 16.
135 Charleton 1654, page 111. See the similar language at page 262.
136 Barrow 1685, 1664 Lectures, page 24. This usage is noted by McGuire and Tamny at *CPQ* 55.

So far we have seen four authors Newton was certainly aware of lay out pieces of the picture sketched at the outset of this section, or, at least, make claims that might have helped Newton to arrive at an understanding of space as a common affection and of common affections as distinct from attributes. The treatments of affections and of space as an affection I have found which come closest to the view I believe Newton held come from two texts that he may have read, but equally well may have not. The first work in question is Caspar Bartholin's *Metaphysica major*.[137] This work was first published in Copenhagen in 1629. But, importantly for our purposes, it was also appended to the end of the 1642 Cambridge printing of Magirus, the edition most likely used by Newton. *Metaphysica major* directly addresses the question "What is an affection?" Bartholin claims that affections are real or are some thing. After all, how could nothing affect something? And he allows that they adhere to the essential nature of their subject. But he denies that affections are the essential nature or property of their subjects. Indeed, he is quite insistent on this point, claiming that if affections were not differentiated in this way, then they would just *be* the subject and therefore cease to be affections. Finally, he goes on to argue that understood in this way, a being's attributes are not affections. The reason is that attributes are not real things, distinct from their subject, in the way that affections are.[138] This final claim rests on a perhaps somewhat idiosyncratic understanding of the attributes of being, one which makes them equivalent to the so-called transcendentals: being, oneness, goodness. But setting this point aside, the view of affections on offer in this text neatly differentiates them from attributes and it allows that they attach to and affect substances, but it nonetheless maintains that they have some level of being that is their own and some determinate nature. Was Newton's view of affections influenced by Bartholin? Certainly there is nothing that proves it was. But it is most likely Newton used the edition of Magirus that contained Bartholin's text. And it is worth noting just how early in the *Metaphysica major* this discussion occurs. Even if the young Newton were just flipping through the first several pages of the text included at the back of his assigned reading, he easily could have come across Bartholin's treatment of affections.

The second text which Newton may or may not have come across that deserves mention is Franco Burgersdijk's *Institutionum Metaphysicarum*. First published in Leiden in 1640 the work—like others written by Burgersdijk—was popular, especially in educational contexts. Cambridge was no exception. As with Bartholin, there is no way to establish with certainty that Newton spent time with Burgersdijk's text or that he was influenced by it. But Newton's account of expenses from his

[137] This is Caspar Bartholin the Elder (1585–1629), not to be confused with his grandson Caspar Bartholin the Younger (1655–1738), a prominent anatomist. The family included a number of notable intellectuals. In connection with Newton, it is also important to differentiate Caspar Bartholin the Elder from his son Rasmus Bartholin, whose work on the double refraction of the Iceland spar was published in the *Philosophical Transactions*. On Rasmus, see Buchwald and Feingold 2013, page 72.

[138] Bartholin 1642, pages 7 and 8; appended to the end of Magirus 1642.

earliest years at Cambridge does contain a tantalizing reference to money paid out in connection with "Burgersdicius."[139] In Book 1, ch. 10 of the *Institutionum*, Burgersdijk identifies a clear (albeit merely rational) distinction between affections relating to *essence* and those relating to *existence*. Affections like unity, truth, and goodness follow from essence. Affections like place and time (and location in place and time) follow from existence.[140] After offering more detailed treatments of the affections of essence, Burgersdijk returns to the affections of existence in chapter 21 and offers claims markedly similar to some made by Newton: "anything that exists, by the very fact that it exists, exists in a place, for that which does not exist in a place, exists nowhere; but that which exists nowhere, truly does not even exist."[141] He then moves on from this claim to establish that a certain infinite space is needed to understand being in place. And he affirms that this space is neither substance, nor accident, but is instead some other real thing. Again, we cannot know that Newton read Burgersdijk, but the resonances are striking.

Whatever Newton's sources were, it is very clear that Newton categorized space as a common affection. Three textual moments deserve attention. The first is the claim in *De gravitatione* that "space is as it were an emanative effect of God and an affection of every kind of being."[142] Newton immediately in the next clause moves to contrast this affection of all being (*omnis entis affectio*) with affections that are *proper* to substances. The implication, then, is that space is not just an affection but a common affection. It can be tempting to focus on the claim that space is as it were an emanative effect of God. But my view is that we should not let this first claim eclipse the second one, the claim that space is an affection of all being. This is especially true in light of the fact that the second claim is unqualified (it lacks the *tanquam*) and in light of the very striking fact, discussed in the previous section, that Newton thinks we have an adequate idea of space apart from God and can conceive of space even while feigning God's nonexistence.

The second key text is also from *De gravitatione*. It comes in a section where Newton is endeavoring to illustrate various properties of space, "not only to show that it is something, but also to show what it is."[143] Here is the part relevant for current concerns:

> Space is an affection of being just as a being. No being exists or can exist which is not related to space in some way. . . . And hence it follows that space is an emanative effect of the first existing being, for if any being whatsoever is posited, space is posited. And the same may be asserted of duration: for certainly both are affections or attributes of a being.[144]

[139] TCL Ms. R.4.48c, 22v.
[140] Burgersdijk 1640, page 59.
[141] Burgersdijk 1640, page 128.
[142] *DG* 36.
[143] *DG* 36.
[144] *DG* 40.

> *Spatium est entis quatenus ens affectio. Nullum ens existit vel potest existere quod non aliquo modo ad spatium refertur... Et hinc sequitur quod spatium sit entis primariò existentis effectus emanativus ^quia posito quolibet ente ponitur spatium^. Deque Duratione similia possunt affirmari: scilicet ambæ sunt entis affectiones sive attributa ...*

There are a number of points to make about this passage.

One place to start is with the fact that Newton again is insisting that space is a common affection. It belongs to or modifies all beings. And it does so just in virtue of their being, not in virtue of any particular feature they have or in virtue of the kind of thing they are. This is consonant with the earlier passage from *De gravitatione* just examined. Admittedly, Newton does use the language of attribute in passing here (*affectiones sive attributa*). But I think this can be discounted somewhat. First, the *sive* functions to equate this particular usage of attribute with affection. Second, the dominant terminology here is clearly that of affections and not that of attributes.[145]

A second point to note is the way that this passage introduces some ontological distance between God and space. The language of emanative effect arises again. But note that this time space is not said to be an emanative effect of *God*, but rather an emanative effect of the first existing thing.[146] On my view, this provides a plausible explanation for why Newton would have used a *tanquam* in the passage considered earlier but did not use one here. Of course, "God" and "first existing thing" will end up being extensionally equivalent for Newton. But the description here matters. I believe that Newton is trying to tell us that space is affiliated with God just, or only, in virtue of the fact that God is a being and not in virtue of any particular features God has or in virtue of God being a special or divine being. Put differently, space is not an affection of God's *nature* but is rather an affection of God's *existence*. So space is only "kind of" or "as it were" an emanative effect of God *qua* God whereas it is more properly said to be an emanative effect of God *qua* first existing being. The reason is that space is an affection of all beings just as beings.

Further support for this interpretation comes from the phrase between carets in the Latin text of the quotation above. This was an interlinear insertion, meaning it was added with a more than normal level of care and intentionality. The standard English translation has it that if any being whatsoever is posited then space is posited. The presence of the *quolibet* is worth stressing. Newton could equally well be

[145] One further point: We can note as well that in the series of drafts for an *Avertissement au Lecteur* mentioned above, Newton refers to space and time as *modes*: "modes of existence of all beings." Koyré and Cohen 1962, page 97; see also page 102. On most taxonomies, the categories of modes and attributes are mutually exclusive insofar as a substance's modes are accidental and can be gained or lost whereas a substance's attributes are immutable. My proposal, again, is that affection is Newton's preferred category and he uses related ones, like attribute and mode, to help his readers triangulate in on his sense of affection.

[146] This point is nicely brought out by the discussion at Janiak 2008, pages 145–146.

saying if *anything at all* or if *whatever object you please* is posited then space will exist. Again, the upshot of this is that the relationship between God and space will not be quite as special as some have thought. Space is a common affection of all beings.

The third passage deeply relevant to this point comes from the very beginning of *Tempus et Locus*. There Newton writes that: "Time and Place are common affections of all things without which nothing whatsoever can exist."[147] With respect to content, this does not add a great deal to the passages from *De gravitatione*. Instead, it serves to reaffirm the basic claims made there. But it is significant given that it shows Newton reiterating those basic claims in a different manuscript, one written at a later date.

Bearing all of this in mind, I feel comfortable concluding that, for Newton, space and time are beings. They are not attributes of any substance and are not attributes of God. While they are ontologically linked to substances and are not fully independent, they have a higher degree of independence than do properties or attributes. And, in part, this seems to be because they are common to or shared between many different substances. This much, it seems to me, is relatively clear from Newton's text.

I freely admit, however, that there are open questions here and that some of them are big and difficult. I have not offered anything like a full theory of affections. I have not made entirely clear the sense in which space and time can be understood as something like (emanative) causal effects or as beings "constituted" by the existence of another being. I am uncertain how to understand space and time's existence as causal effects or ontic consequences in addition to or in connection with the status they have as common affections. And I think the way in which space and time are common to, or are shared between, multiple different entities presents some deep puzzles. In this sense, there is room for further research. My hope is just that this research can be conducted under the rubric of space and time as common affections, rather than under the rubric of space and time as divine attributes. For now, however, given that the focus of this book is primarily on Newtonian *substances*, rather than Newtonian affections, enough has been said about space and time.

3.8. Conclusion

This chapter and the one that preceded it have attempted to articulate some basic Newtonian metaphysical commitments relating to God. Specifically, they examined Newtonian arguments for God's existence, issues surrounding necessity and

[147] McGuire 1978b, page 117.

choice, the architecture of the divine substance, and God's relationship to space and time.

Of course, not all questions here could be answered. Newton is often at pains to stress the high degree of freedom God has in creating and in acting. This, combined with the fact that our only (non-Scriptural) access to God is through divine works, can make it difficult to work backwards and reverse engineer a complete picture of what God is like. But this need not be troubling; here it is worth remembering that, from Newton's perspective, what matters is God's moral authority and monarchical control over creation, and not a full and complete accounting of the divine ontology. We should strive to understand God by examining nature. The General Scholium tells us that "to treat of God from phenomena is certainly a part of natural philosophy."[148] But we should be prepared for the fruits of this striving to be primarily practical, rather than theoretical. As Query 31 expounds the point:

> And if natural philosophy in all its parts, by pursuing this method, shall at length be perfected, the bounds of moral philosophy will be also enlarged. For so far as we can know by natural philosophy what is the first cause, what power he has over us, and what benefits we receive from him, so far our duty toward him, as well as that towards one another, will appear to us by the light of nature.[149]

[148] *Principia* 943. The second edition has "experimental" for the third's "natural" in this sentence.
[149] *Opticks* 381.

4
Newton's Theory of Bodies

4.1. Introduction

What are Newtonian bodies? One might look to the *Principia* for an answer. Newton there defines quantity of matter as "a measure of matter that arises from its density and volume jointly."[1] He then notes that this is the quantity he has in mind "whenever I use the term 'body' or 'mass' in the following pages."[2] This definition is good so far as it goes, and we will have reason to return to it in subsequent chapters. But for those interested in the deep metaphysics of Newtonian material substance, it does not offer very much. Fortunately, there are other sources. My view is that examining these other sources reveals that Newton developed a theory about the nature of bodies that was bold, innovative, and—by the standards of the period—quite compelling. The core of this theory is found in *De gravitatione*. This chapter will offer a quick characterization of that theory and will trace its development prior to *De gravitatione* as well as identify moments in Newton's later career where the theory seems to be playing some role. The next chapter will develop a much more detailed interpretation of the account in *De gravitatione*, focusing in particular on the relationship between God and bodies. The chapter following that will focus on developments in the theory after the period of *De gravitatione*.

4.2. *De gravitatione*: A Neutral Summary

Many of Newton's texts are relevant for thinking about the metaphysics of bodies. But, far and away, the most important is *De gravitatione*. It is the text that forms the backbone for several other interpretations of Newtonian bodies. And it will be the backbone of my interpretation as well. Therefore, a quick overview of Newton's claims about bodies in *De gravitatione* will be useful before proceeding further. Of course, in the end it will be the fine-grained details of the account, in combination with other texts, that will determine Newton's precise position. But for the present moment, the broad contours will be helpful for understanding some of the key

[1] *Principia* 403.
[2] *Principia* 404.

Newton's Metaphysics of Substance. Patrick J. Connolly, Oxford University Press. © Patrick J. Connolly 2025.
DOI: 10.1093/9780191959240.003.0004

issues and also for understanding how and why various divergent interpretations of the text have come about. The overview below is intended to be "neutral" in the sense that it tries to avoid a description of the account that favors any one interpretation over another. This attempt is no doubt only partially successful. Those who worry the description shows signs of prejudice, in one way or another, can rest assured that detailed analysis of both the text and the various interpretations of it on offer will be provided below and in the next chapter.

The account begins with God and absolute space. Newton first attends to the impenetrability of bodies. He allows that it is surely within God's power to make it such that a region of space excludes bodies from entering into that region. If God were then to make it the case that the impenetrability in this region of space could move in various determinate ways while retaining its quantity and shape, then we would have something decidedly body-like. Finally, we could imagine that these things were capable of interacting with our minds, of producing ideas in them, and of being moved by the exercise of our wills. Newton suggests that this account of bodies does away with the need for prime matter into which the form of a body can be impressed or for a substratum in which the properties of body can subsist or inhere. His suggestion seems to be that bodies are nothing over and above a certain small number of divinely-created powers.

4.3. The Status of the Theory in *De gravitatione*

The previous section, in attempting to give a neutral summary of the account of bodies in *De gravitatione*, noted that there is considerable disagreement over the *content* of the views on bodies Newton puts forward in that text. But there is also uncertainty around the *doxastic status* of the views there. These questions surrounding Newton's degree of commitment to the views in *De gravitatione* deserve to be addressed before considering the views themselves. At one end of a spectrum, one might hold that Newton was firmly committed to God having created bodies in just the way described in the manuscript and further that he maintained this view consistently throughout his mature career. At the other end of the spectrum, one might hold that the creation story in *De gravitatione* is little more than an intellectual exercise or intuition pump and that it merely represents Newton thinking through some relevant issues while developing a series of anti-Cartesian arguments.

While one might be hard pressed to find representatives of either of these extreme positions in the literature, many of the positions between these two poles are occupied. Christopher Conn and Martin Tamny, for example, are inclined to see Newton as straightforwardly accepting the account.[3] A weaker version of this

[3] Conn 1999, page 438; Tamny 1976, page 182; Tamny 1979, page 50.

can be found in Alan Gabbey, who suggests that Newton might have believed this to be the correct account of creation but would have considered it to fall short of knowledge.[4] Weaker still might be the view of Catherine Wilson, who claims the account is "lightly agnostic."[5] Toward the other end of the spectrum are authors like Howard Stein; he claims that the account is "fundamentally conjectural."[6] Other commentators who take a cautious approach to the issue and emphasize the sense in which it might constitute mere speculation include Richard Westfall, Benjamin Hill, Lisa Downing, and Rob Iliffe.[7] Elliot Chen also categorizes the account as speculative and claims that it only means to address a narrow set of issues and is primarily targeted at criticizing Descartes.[8] A position in the middle is carved out by Zvi Biener and Chris Smeenk who make it clear that Newton shied away from affirming anything about the nature or essence of bodies but engaged in "speculation" about bodies in order to provide a theory that was "*sufficient* to serve as a basis for physical theory and *sufficient* to capture the phenomenal properties of bodies."[9]

This uncertainty with respect to how much faith Newton had in the *De gravitatione* account of bodies is reflected in the secondary literature's uncertainty over the correct way to refer to the account. Howard Stein refers to it as the "creation story" or the "creation hypothesis."[10] Andrew Janiak also opts for "creation story."[11] Ed Slowik calls it the "determined quantities of extension hypothesis" or "DQE hypothesis."[12] Other authors also refer to the account as a hypothesis.[13]

Much of the speculation on this issue is a result of Newton's own framing of the creation theory. Surely all interpreters of *De gravitatione* must at least acknowledge that he offers a great deal by way of preface before plunging into his exposition of body. Here is the relevant paragraph:

> [I]t remains to give an explanation of the nature of body. Of this, however, the explanation must be more uncertain [than that given of extension], for it does not exist necessarily but by divine will, because it is hardly given to us to know the limits of the divine power, that is to say, whether matter could be created in one way only, or whether there are several ways by which different beings similar to bodies could be produced. And although it scarcely seems credible that God could create beings similar to bodies which display all their actions and exhibit

4 See Gabbey 2011, page 435. See also the linguistic considerations at footnote 15.
5 Wilson 2019b, page 434.
6 Stein 2002, page 274.
7 Westfall 1971, pages 339–340; Hill 2003, footnote 5; Downing 2014, page 106; Iliffe 2017, page 103.
8 Chen 2020.
9 Biener and Smeenk 2012, page 111. Italics in original. See also, however, the more skeptical tone at page 117 and at Biener 2020, pages 186–187.
10 Stein 2002.
11 Janiak 2008, page 146.
12 Slowik 2009.
13 McGuire 1982, page 147; Kochiras 2013, pages 343–344; Downing 2014, page 106.

all their phenomena, and yet would not be bodies in essential and metaphysical constitution, as I have no clear and distinct perception of this matter I should not dare to affirm the contrary, and hence I am reluctant to say positively what the nature of bodies is, but I would rather describe a certain kind of being similar in every way to bodies, and whose creation we cannot deny to be within the power of God, so that we can hardly say that it is not a body.[14]

There is certainly material here that could support positions toward either end of the spectrum sketched above. God's tremendous power and our limited understanding of it support a very cautious approach to any understanding of bodies we might devise. So the account could be read as strictly hypothetical or exploratory. On the opposing side, Newton does say that God certainly could create bodies in this way and that they would be indistinguishable from bodies as we know them in every way.

One thing that might shed some light on Newton's apparent hesitancy here is the fact that Descartes had displayed a similar sort of hesitancy in similar contexts.[15] Perhaps most importantly, Part 3 of the *Principles*, which addresses the "Visible World," begins with an admonition to consider our own limits especially with respect to the "infinite power and goodness of God" and to therefore reason very cautiously and carefully about such matters.[16] This neatly matches Newton's claim that we do not know the limits of divine power. And in sec. 44 of Part 3, Descartes asks that his views be considered as hypotheses but claims that even so they will be "worthwhile if everything deduced from [them] agrees with our observations; for if this is so, we shall see that our hypothesis yields just as much practical benefit for our lives as we would have derived from knowledge of the actual truth."[17] This seems to track Newton's thought that the "bodies" he describes would be, from our perspective, empirically identical to the bodies we observe in nature; they would display all the actions of bodies and exhibit all their phenomena and would be similar in every way to bodies. More generally, when Descartes speaks of how God might have created the universe in the *Discourse on Method*, he is quick to emphasize that his theory is not meant to explain exactly how God *did* create the universe but rather one way in which God might have done so.[18] This comparison to

[14] *DG* 41–42. While less commented on, there is also a passage at the end of *DG*'s discussion of bodies which expresses a similar sentiment. *DG* 47–48.

[15] This similarity is briefly commented on at Stein 2023, page 133 and Wilson 2019b, pages 434–435.

[16] Principles, Part 3, secs. 1–2. *CSM*, I, 248.

[17] *CSM*, I, 255.

[18] Of course, Descartes' reasons for hesitancy here are slightly different than those of Newton in *DG*. Descartes was worried that his account might fall afoul of orthodox Catholic interpretations of the creation story in Genesis. Newton ultimately endorsed a different strategy for reconciling Scriptural accounts of creation with philosophical ones. See *NCS* #247 as well as Snobelen 2008, Janiak 2012, and Janiak 2015. And, in any event, Newton was operating in a more tolerant context and certainly unconcerned with what Catholic authorities might think of him and his views. According to Conduitt, "Sir I[saac] had the happiness of being born in a land of liberty & in an age where he {might} speak his mind—not afraid of {the} Inquisition as Galileo was for {saying} the sun stood still & the earth {moved}

Descartes might give us some reason to discount Newton's apparent hesitancy. For all his professed modesty, Descartes clearly had no reservations about whether or not his views were the correct ones. Something like this might very well be true of Newton. In the same way that Newton casually appropriates the familiar language of clear and distinct perception in the passage under consideration, the general notes of caution he sounds here might have been another effort to merely deploy a common literary trope.[19]

In keeping with this thought, my view falls toward the end of the spectrum that claims Newton took the account of body's creation in *De gravitatione* very seriously. In the first instance, I think commentators have been too quick to read the paragraph quoted above as evidence that Newton took himself to be speculating or offering something purely hypothetical. Even setting aside the thought just canvased, that Newton may have been merely echoing an approach employed by Descartes, the text itself can be seen as quite assertive. Newton does allow that God's power outstrips our ability to understand it. But he also says that, at least from our epistemic perspective, "it scarcely seems credible" that God could create beings that display all the phenomena of bodies but which were not "bodies in essential and metaphysical constitution."[20] Newton is very clear that he believes the entities he describes in *De gravitatione* would, in fact, display all the phenomena displayed by bodies. So we are entitled to read him as thinking that it is in no way credible to see them as not essentially bodies. Indeed, this is just where the paragraph ends, with an assertion that it is not at all possible for us to say with certainty that these entities are not bodies. Read this way, it seems like Newton's basic attitude toward the account he gives is one of acceptance. It is true that Newton notes we cannot be certain the theory is correct. What I want to emphasize is that he is equally insistent that we cannot be certain the theory is wrong and also have no reason for thinking it is. Put differently, Newton seems to be suggesting that while

his works not in danger of being expunged as DesCartes's was nor he obliged to go into another country as Descartes was into Holland to vent his opinions, nor reduced to the miserable shifts as Descartes was of saying his Philosophy was the Philosophy of Moses & that he could proue Transubstantiation mathematically." Keynes Ms. 130.07, 5r–5v.

[19] Two further considerations: First, the language of clarity and distinction is characteristically Cartesian, but it is worth noting that it also had a more widespread use in the period, even among anti-Cartesians. See Malcolm 2002, pages 181–182. Second, a comparison to Locke might be helpful here. Much has been made of his claim in the *Essay's* "Epistle to the Reader" to be a mere underlaborer to Newton and the other great natural philosophers of his era. Locke 1975, pages 9–10. Scholars are right to give the passage attention and it is surely revealing. But the fact that very similar claims can be found in many other texts from authors in and around the early Royal Society should give us reason to avoid placing too much emphasis on Locke's claims. See, e.g., Hooke 1665, unnumbered pages in Preface; Sprat 1667, pages 28–29; Boyle 1999–2000, volume 2, page 21; Glanvill 1668, page 91; Power 1664, page 192.

[20] *DG* 41. Indeed, Newton's Latin (*haud credible videtur*) admits of more forceful English translations than "scarcely seems credible." It could equally well be rendered as saying it seems "not at all credible" or seems "by no means credible."

we ought not dogmatically affirm the theory, we should—pending the advent of new evidence or considerations—accept it.

I have just claimed that even the key text commentators have identified as evidence that Newton was hesitant about the account of bodies offered in *De gravitatione* is, when properly interpreted, evidence that he actually had a high degree of confidence in that account. My other reason for seeing Newton as genuinely committed to the account, rather than considering it as a mere hypothesis, stems from the fact that it seems to enjoy considerable longevity in his thought. The next several sections are geared at showing that there are instances in which Newton either outlines or gestures toward what is either the same or a very similar account at several points in his career, including many that post-date the composition of *De gravitatione*.[21] The approach to bodies in *De gravitatione* is highly sophisticated. It clearly took some work to develop and describe. It also offers a compelling picture for someone with Newton's epistemological and theological commitments. So I think it is unlikely that Newton would have developed a theory like this without a high degree of investment in it being correct. And nothing in his later career gives evidence that he might have changed his mind about the strength of the theory or that he might have abandoned the fundamentals of the theory.

4.4. Origins of the Account in *CPQ*

It is abundantly clear that thinking about Descartes is key for any appropriate interpretation of *De gravitatione*. As already noted, *CPQ* shows signs of Newton's deep engagement with Descartes during a formative period. And the fact that what Newton offers in *De gravitatione* is in many ways designed to be a *rejection* of Cartesian views hardly makes Descartes less important.[22] After all, at one point, Newton says that the overthrow of Cartesian views on matter and extension is the "most important" task for those hoping to lay true foundations for natural philosophy.[23]

With that being said, it is also important to remember that *De gravitatione* contains important *positive* proposals and this means that the document is much more than a rejection of or response to Descartes. This is, of course, true with respect Newton's views on absolute space and time in the text. But it is also true with

[21] This view runs counter to one put forward at Henry 2011, page 23: "while the discussion of space in *De gravitatione* shows marked and detailed similarities with other pronouncements and discussions of space from different parts of Newton's career, the discussion of body is a unique one-off. . . . As far as I know, Newton, who was given to recycling and re-presenting ideas throughout his career, never discussed such an idea of body anywhere ever again." A more nuanced, but essentially similar, view is at Jalobeanu 2007, page 94. A contrasting view is available at Iliffe 1995, page 451.

[22] As McGuire 1978b, page 124 puts the point: "At every turn, Newton's characteristic theories are developed in opposition to Descartes."

[23] *DG* 35.

respect to Newton's views on the nature of body. And, here, at first glance, it seems that Newton's views are strikingly original. It is worth, therefore, considering where they might have come from.

CPQ's discussion of the atoms that comprise "first matter" contains a section designed to clarify what these atoms might be. Specifically, it is designed to "help the conception of the nature of these leasts, how they are indivisible, how extended, of what figure, etc."[24] The clarification begins with mathematical points. We are asked to suppose "a number of mathematical points were imbued with such a power as that they could not touch or be in one place."[25] This should be read in light of Newton's earlier rejection of the idea that mathematical points *themselves* could be first matter. The problem with that supposition was that mathematical points lacked sufficient reality to constitute bodies. Here is that analysis in full:

> [W]hat want dimensions cannot constitute a body in their conjunction because they will sink into the same point. An infinite number of mathematical points sink into one, being added together, and that, being still a mathematical point, is indivisible. But a body is divisible.[26]

Bodies have to be divisible, but mathematical points are indivisible. If mathematical points could be stacked alongside one another, then we would have divisibility because we could draw lines between them. The problem with trying to aggregate mathematical points in that way, however, is that they will "sink" into one another. There is nothing to keep them from coming together. They have nothing in them that prevents co-location.

Thus, Newton's solution is to not start with *bare* mathematical points. Rather, he suggests starting with mathematical points that have a certain power imbued into them. Specifically, the power to avoid contact and thereby avoid co-location. Beings like these are capable of being aggregated in a way that mere points are not. This happens by bringing them as close together as is possible. The power that they have to not touch will ensure that they do not "sink" into one another and also that the absolute minimal possible distance obtains between them. Putting several of these points in a line will produce extension because, on pain of the points touching, there will of necessity be some distance between the different parts of the line. Yet the individual points themselves will be indivisible because, *ex hypothesi*, they are mere points.[27]

[24] *CPQ* 343.
[25] *CPQ* 343
[26] *CPQ* 337.
[27] McGuire and Tamny have suggested that some of Newton's thinking here might be indebted to Francis Line's *Tractatus de corporum inseparabilitate* (Line 1661, pages 170–182) or to Boyle's discussion of Line's views in his *A Defence of the Doctrine touching the Spring of the Air* (Boyle 1662, pages 101–107). *CPQ* 342, note 13. Given the amount of Boyle Newton was reading at the time, it is certainly plausible that he may have encountered this discussion by the time he wrote this entry. See Tamny 1976,

Newton's goal in this section is to illustrate how there might be true physical minima: parts of matter so small as to be indivisible even in thought. Whether his account is successful in establishing this possibility is up for debate.[28] Either way, the account is interesting because it foreshadows the *De gravitatione* theory of bodies. There are, of course, some key dissimilarities. First, and perhaps foremost, is the fact that in *CPQ* powers are located at mathematical points. In *De gravitatione* they will be located in extended regions of space. But in *CPQ* Newton had not yet developed the account of absolute space that he appealed to in *De gravitatione*. This is particularly important because absolute space, on Newton's mature theory, is an entity or a being in a way that mathematical points are not. As Newton himself points out, a mathematical point is ontologically nothing and therefore not "capable of powers of qualities."[29] This means that the account in *CPQ* is merely an intuition pump or a cognitive device for letting us think about atoms as true minima.[30] By contrast, the *De gravitatione* creation story is meant to offer a genuinely possible account of how bodies might have been brought into being. Second, the mathematical points are endowed with only one power whereas *De gravitatione* discusses the possibility of multiple different powers being co-located. And third, there is no suggestion in *CPQ* that God might be responsible for causing points to have these powers and, indeed, no suggestion at all about how the powers might come to exist.

Even allowing for these important differences, the structural similarity between this part of *CPQ* and the creation story in *De gravitatione* is notable. Both accounts are aimed at offering a story that explains fundamental features of bodies. Both accounts begin with a concept that is fundamentally geometrical. And both accounts envision the situating of powers with respect to the geometrical entity. The power at issue in *CPQ*, variously described as a "power as that they could not touch all over and be in one place" and the "power to resist conjunction with another"—is

ch. 8. I am skeptical, however, that there is much by way of meaningful similarity between these texts and Newton's entry. Line does not seem to identify being extended with being divisible. But this identification seems central to Newton's argument. If a connection between Newton and Line could be established on this point, however, it would be of interest given their subsequent disputes over optics. For more on Line see Reilly 1969. I am more inclined to think that Newton's approach here follows certain lines of thought in Henry More's work. See Reid 2012, pages 63–73. Another potential source is Galileo; there are some resonances with material from the "First Day" in the *Discourse on the Two New Sciences*. See De Angelis 2021, pages 20–33, especially page 24.

[28] See Janiak 2000, sec. 2. In connection with this it is worth noting that much of this entry is marked for cancellation. Janiak argues, plausibly, that we ought not make too much of this. But the cancellation may indicate that Newton was ultimately unsatisfied with details of the account here and therefore was motivated to come up with the distinct, albeit closely related, theory in *DG*.

[29] *CPQ* 423.

[30] Thus Newton writes that "though it be impossible that the thing [points endowed with powers] should be so, yet it is not so to conceive it; nay, if atoms be so small it is necessary to conceive they are bounded and touch others by mathematical points and superficies at so small a distance as is described here." *CPQ* 423–425.

clearly akin to the first power mentioned in *De gravitatione*.[31] The bodies in that later text have the power such "that two of this kind cannot coincide anywhere, that is, that they be impenetrable ... "[32] Given all of this, I think it is clearly possible to see the germ of the *De gravitatione* creation theory in *CPQ*.

4.5. Early Optical Papers

There are also hints of something like the *De gravitatione* account of bodies in Newton's early optical papers. The relevant texts date from late 1675 and early 1676. On 7 December 1675, Newton sent a letter to Oldenburg containing some views on the nature of light. It included the following lines: "Perhaps the whole frame of Nature may be nothing but aether condensed by a fermental principle. Thus perhaps may all things be originated from aether."[33] On its own, this is perhaps suggestive, but rather underwhelming. On 25 January 1676, however, Newton wrote again to Oldenburg asking him to make a change.[34] The amended passage was to read as follows:

> Perhaps the whole frame of Nature may be nothing but various Contextures of some certaine aethereal Spirits of vapours condens'd as it were by praecipitation, much after the manner that vapours are condensed into water or exhalations into grosser Substances, though not so easily condensable; and after condensation wrought into various forms, at first by the immediate hand of the Creator, and ever since by the power of Nature, wch by virtue of the command Increase & Multiply, became a complete Imitator of the copies sett her by the Protoplast. Thus perhaps may all things be originated from aether.[35]

The revised passage is far more interesting.

It does have to be acknowledged that this conception of the origin of bodies diverges significantly from the one on offer in *De gravitatione*. The register of the

[31] *CPQ* 343 and *CPQ* 423. It is interesting to compare the characterization of the powers given to mathematical points here with the formal definition of force offered in "On Reflection." That text, written shortly after the period of *CPQ* and concerned with collisions or impact laws, has it that "Force is the pressure or crouding of one body upon another." CUL Ms. Add. 4004, 10v. Also interesting in this connection is a passage from Hooke 1678, page 8, mentioned in Kochiras 2013, note 21: "Body is somewhat that doth perfectly fill a determinate quantity of space or extension so as necessarily to exclude all other bodies from being comprehended within the same Dimensions. [¶] I do therefore define a sensible Body to be a determinate Space or Extension defended from being penetrated by another, by a power from within." That said, while Hooke's phrasing has some resonance with Newton's, in what follows Hooke goes on to characterize a conception of exclusion or impenetrability quite different from the one developed in *DG*.

[32] *DG* 43.

[33] *NCS* #146.

[34] *NCS* #153.

[35] *NCS* #146.

text is much more reminiscent of ones like *Of Natures obvious laws* and *De aere et aethere*.[36] And the invocation of an aether would seem to put some distance between this text and the resolutely vacuist *De gravitatione*. But while it is important not to downplay the differences between the texts, my view is that there are also similarities that are well worth noting.

This is a theory about creation and the origin of bodies; it offers a description of how "all things" may be "originated." And it also invokes the creative activity of God. The immediate hand of the Creator is what, at first, gives shape to the bodies. Admittedly, whereas the raw materials for bodies in *De gravitatione* are space and powers, this account makes appeal only to the aether and a fermental principle. But it is worth highlighting that the aether is fundamentally spatial and that the described act of condensation involves a sort of reconfiguration into discrete parts of space, as does the process of the condensed aether being "wrought into various forms."[37] In this connection, it is worth noting that some interpreters have even seen fit to describe the bodies in *De gravitatione* as being created through the "thickening" of space.[38] Again, it is important to neither overstate the similarities nor to skip over the dissimilarities, but there are certain resonances between Newton's claims about the origin of bodies in this text and his treatment of the topic in *De gravitatione*.

The same is also true for some of what Newton says about light later on in the text. Here is the relevant passage with letters added in brackets to facilitate discussion:

> In the fourth place therefore I suppose Light is neither this Aether nor its vibrating motions, but something of a different kind propagated from lucid bodies. [A] They that will may suppose it an aggregate of various peripatetic qualities. [B] Others may suppose it multitudes of unimaginable small & swift Corpuscles of various sizes, springing from shining bodies at great distances one after another . . . [C] God who gave Animals self motion beyond our understanding is without doubt able to implant other principles of motion in bodies wch we may understand as little. Some would readily grant this may be a Spiritual one; yet a mechanical one might be showne, did not I think it better to passe it by.[39]

Newton invites us here at [A] to imagine a kind of being that is nothing beyond an aggregate of various qualities. This is striking insofar as there is no suggestion of any thing or substance that instantiates or supports the qualities. The aggregate

[36] Though see Henry 2023, page 250, for an important point about the differing conceptions of aether across these texts.

[37] There is an interesting parallel here to Newton's discussion of creation in his correspondence with Thomas Burnet. See *NCS* #247.

[38] See Bennett 2005, page 181 and Nuovo 2017, page 171.

[39] *NCS* #146.

alone is sufficient to constitute light. At [B] we are offered an alternative conception; light could be particles or corpuscles moving in various ways. But in order for the sense of the passage to work, it must be the case that either conception is sufficient and that both will allow readers to converge on the same thing. In this sense, the two conceptions are interchangeable with one another. Put differently, this text opens up the possibility of an approach to particles that understands them *as* aggregates of powers or qualities.

Also notable in the passage is the point at [C] about the principles underlying or explaining the motions of these particles. First, there is a claim about *what* was done. Specifically, God implanted certain powers or principles in bodies. These implanted powers, following God's initial act, presumably remain in the bodies and account for various phenomena they display. Second, there is a claim about the *manner* in which God does this. And Newton's suggestion here is that God's method might defy our attempts to understand it. Both of these claims have thematic parallels in *De gravitatione*.

4.6. To Locke, Pembroke, and Coste

According to one of the notes Pierre Coste included in his French translation of Locke's *Essay*, Newton shared his theory of the creation of bodies with John Locke and the Earl of Pembroke and this provides the explanation for an otherwise cryptic passage added to the second edition of the *Essay*.[40] This conversation must have happened some time between Locke and Newton's meeting, which occurred in late 1689 or early 1690 shortly after Locke's return from the Netherlands and the publication of the second edition of the *Essay* in 1694. And, on Coste's telling, Newton was happy to share the theory again with Coste himself "many years after Locke's death."

There is little reason to doubt Coste's veracity here. Put differently, the note is important because it shows that Newton seemed to think some version of the *De gravitatione* theory was good enough to share with others—Locke and Pembroke—in the early part of the 1690s.[41] And the broad outlines of Coste's description make

[40] Locke 1729, page xliv. Full details are given in Bennett and Remnant 1978. McGuire 1968a, page 193; McGuire 2000, pages 290–291; Tamny 1976, pages 181–182; Janiak 2008, page 125; and Dempsey 2014, footnote 38, also reference the importance of Coste's note as evidence of Newton's continued commitment to the theory outlined in *DG*.

[41] There may be, however, some exceedingly slight reason for doubt. The 1729 edition of Coste's translation of the *Essay* appeared after Newton had died. This edition has it that the relevant conversation was between Newton, Locke, and an unnamed English lord. The 1735 edition of the translation identifies the English lord as Pembroke. See Locke 1735, page 531. Suspicious minds might wonder about the fact that Coste occluded the name of the one person who could vouch for the story while that person was alive but included it after he had died. On a different note, Bennett and Remnant (footnote 7) somewhat puzzlingly write that the 1735 edition offers a marginal note identifying the lord as "the late Count of Penbrocke, who died in February of the present year, 1738." I consulted three digitized copies of the relevant work (one held by the National Central Library in Rome, one by the Austrian

it all but certain that Newton communicated some version of the theory to him some time after Locke's death in 1704. Beyond this, however, I think we should be very hesitant about seeking to reconstruct Newton's views in any detail on the basis of Coste's report. This is for a number of reasons.

First, it is unclear how long the gap was between Coste's conversation with Newton and his recounting of that conversation in the translation of the *Essay*. Coste merely remarks that he spoke to Newton long after Locke's death. Given that Locke died in 1704 and Coste first added the note to the 1729 edition of the translation, it is possible that Coste was remembering a conversation that had happened nearly twenty years in the past. Of course, the gap between the two events might have been much smaller.[42] But, however big the gap was, it is certainly possible that Coste may have forgotten or incorrectly remembered some details in his later recounting.

Second, there are numerous differences between, on the one hand, Coste's version of the account and, on the other, the versions of the theory offered by Newton in *De gravitatione* and the *Opticks* text discussed in the next section.[43] Some of these are minor. Coste has it that impenetrability is due to God making it so that nothing could enter (*rien ne pût entrer*) a certain portion of space. But in other accounts, Newton specifies that it is merely the prevention of other bodies or of other quantities of extension endowed with a similar power of impenetrability. This is important as Newton believed that spaces occupied by bodies could additionally be occupied by God and finite thinking things.[44] Coste also has Newton discussing *essential* features of bodies, something he was normally hesitant to do. Other differences are far more significant. Mobility, which on Newton's versions of the theory is considered to be its own, distinct, power is on Coste's telling a product of God communicating the impenetrability to a different region of space.[45] More

National Library, and one by the National Library of France). These editions had not a marginal note, but an asterisked note at the bottom of the column which clearly refers to "Pembroke" and correctly gives his year of death as 1733. Bennett and Remnant may have been working from a photocopy or microfilm which reproduced the text unfaithfully.

[42] Coste had translated some of Newton's works in 1714 and ultimately produced the first French translation of the *Opticks*, published in Amsterdam in 1720. If Newton and Coste were in touch at this time, then the gap between Newton's recounting of his conversation and Coste's reporting on it would be roughly a decade. For details of Coste's translation of Newton, see Baillon 2009.

[43] In this sense, I think it is going too far to say, as Dempsey does, that "the account related by Coste is essentially the one found in *De gravitatione*." See Dempsey 2006, footnote 9 and also Dempsey 2014, page 99.

[44] It is also important for avoiding the worries about circularity raised at Bennett and Remnant 1978, page 7. See the excellent discussion of this point at Downing 2014, pages 106–110. On the flip side, this stipulation makes Newton's version of the view subject to problems of "substance counting" in a way that Coste's version is not. See Kochiras 2009 and Kochiras 2011.

[45] This point is important for the second set of worries raised at Bennett and Remnant 1978, page 7. More generally, it seems to me that the low opinion that Bennett and Remnant have of Newton's theory stems, in large part, from the fact that much of their interpretation begins with Coste's account rather than with *DG* itself.

generally, Coste's account does not speak of *powers* had by bodies or located in spaces but rather makes God the sole agent. As will be emphasized later, this marks a major departure from Newton's own way of describing things, one with serious implications. A further discrepancy, one unrelated to the content of the theory, has to do with the time of its development. Coste reports that Newton had devised the theory on the day of his conversation with Locke and Pembroke. But the similarities between what Newton told Coste and *De gravitatione*, which must pre-date the conversation between Newton and Locke, are so striking that this seems implausible. So either Newton misled Coste about the origin of the theory, Coste fabricated this particular detail, or there was some miscommunication between Newton and Coste.[46]

Third, there are some general doubts about Coste's reliability and philosophical skill. The source of these doubts is Locke himself.[47] He seems to have had a surprisingly low opinion of Coste, often treating him in a curt manner. For example, he left Coste nothing in his otherwise quite generous will. Intellectually, Locke was often dismissive of Coste's questions and efforts to engage him. And, upon Molyneux's inquiry, Locke painted a fairly negative picture of him, allowing that he was ingenious but saying of his skill in mathematics and natural philosophy that "I think it is not much."[48] In context, Locke's suggestion was that Coste was suitable as a tutor for the young Francis Masham and perhaps as a translator, but not for other tasks. Other of Coste's contemporaries seem to have had a similarly low opinion of him.[49] We can note that even in the passage from Book IV, ch. 10 of the *Essay* that is under consideration, Coste seemed to misunderstand Locke's use of "in this place." Locke had intended to refer to a particular part of the *Essay* but Coste's translation gives a geographical sense of the phrase taking it to refer to "this place on Earth."[50]

In summary, Coste's note is important because it provides further evidence that Newton endorsed the theory of bodies from *De gravitatione* later in his career and was even willing to share the theory with select others. But, certain concerns about Coste's account make it unsuitable to do load-bearing work in a precise reconstruction of the theory itself.

[46] Stein 1970, pages 274–275, note 11, is the only commentator I have encountered who thinks it possible *DG* might have been written as late as the 1690s and that, consequently, Coste might be accurately reporting something Newton truthfully told him.

[47] The following is drawn primarily from Milton 2008, sec. 1. But see also the rather absurd claim Coste reportedly made about a conversation he had with George Berkeley in the *Mémoirs de Trévoux*, March 1750, Article 37, page 677.

[48] Locke 1976–1989, volume 6, page 294. For the context of Locke's remarks, see Molyneux's questions at pages 133–134 and 220.

[49] See the discussion in Baillon 2009, pages 1–2.

[50] I take this point from Bennett and Remnant 1978, footnote 2.

4.7. Query 31

The most significant text apart from *De gravitatione* for these issues comes from Query 31 of the *Opticks*. Here we get a version of the theory that is simultaneously somewhat veiled and developed in much greater detail.[51] It is veiled with respect to the specifics of God's activity and God's situating of powers in space. But it is more detailed with respect to the powers located. Here is the first portion of relevant text:

> All these things being consider'd, it seems probable to me, that God in the Beginning form'd Matter in solid, massy, hard, impenetrable, moveable Particles, of such Sizes and Figures, and with such other Properties, and in such Proportion to Space, as most conduced to the End for which he form'd them.[52]

Some may question the extent to which this really does represent a continuation of the strain of thought about the creation of bodies outlined in *De gravitatione*. As was said, Newton is less forthcoming about the details of God's activity than he is in the earlier text. Nevertheless, the parallels are substantial. Here, Newton is offering an account of what happened "in the beginning," that is to say an account of how God created bodies. And his emphasis that this particular creation process "seems probable" is highly reminiscent of the epistemic caveats that preface the theory in *De gravitatione*. In the *Opticks*, God creates bodies by ensuring that there are particles constituted by certain powers and those powers are in specific proportions to space. This is not very far from the situating of powers in determined quantities of space discussed above. Further, impenetrability and mobility are familiar from *De gravitatione*. The sentences that follow define solidity in terms of impenetrability and hardness in terms of mobility. As the passage goes on, solidity is characterized with respect to non-porousness. It indicates something about the way impenetrability is distributed in space. A particle shaped like a doughnut would not count as solid on Newton's definition, but a cube or spherical particle, because its impenetrability is distributed contiguously through space, would. Hardness is characterized in a way that can be understood with reference to the power of mobility in *De gravitatione*. Newton describes it as ensuring that the impenetrable region will never break into two impenetrable regions (at least through any natural process). And it also entails that the region will never become deformed or alter its spatial distribution. So hardness indicates something about how the impenetrable region will always move in a way that ensures its continuity of size and shape.[53] So, in this sense, we have it that God created things defined by their impenetrability and

[51] Stein 1990, pages 36–37, has noted the resonances between these texts and offers a similar analysis.
[52] *Opticks* 375–376. See also the similar language on pages 377–378.
[53] Compare this with the characterization of hardness at *DG* 53.

mobility and the way their impenetrability and mobility are related to space over time. This represents clear continuity with the line of thought in *De gravitatione*.

Of course, Query 31 also reflects some changes, changes that I believe constitute positive developments. Perhaps most obvious in the passage above is the inclusion of *massy* on the list of defining characteristics for the originally created bodies. More importantly, Newton writes that in addition to the characteristics he mentioned, these original bodies were created by God "with such other properties" as conduced to God's purposes. This is important because it introduces a great deal of flexibility into the account. These changes will be discussed in Chapter 6.

4.8. Disputes with Leibniz

In the context of his later career, some of Newton's clearest statements on metaphysical issues come as a result of his engagement with the thought of Leibniz. It is important to recognize that these are polemical statements. Newton is always trying to show that his views are superior to Leibniz's. But this need not diminish their value for current concerns. In some sense, the opposite is true. The fact that Newton is so often highlighting some contrast between his positive views and a reconstruction of a Leibnizian position makes these claims useful for discerning not just Newton's views but also his reasons for being committed to them and what he saw as their attractive features.

A first hint of the *De gravitatione* theory in Newton's dispute with Leibniz can be found in the *Account of the Commercium Epistolicum*. In typical fashion, Newton offers a list of the competing views he and Leibniz take on a position and it is meant to be obvious which of the two is superior. Here is the entry relevant to the account of bodies: "The one (by way of enquiry) attributes it to the power of the creator that the least particles of matter are hard: the other attributes the hardness of matter to conspiring motions, and calls it a perpetual miracle if the cause of this hardness be other than mechanical."[54] What is interesting here is Newton is clearly allowing that, with respect to the least particles of matter (the fundamental bodies), he endorses a non-mechanical account of hardness. Leibniz's claim that it can be a result of conspiring motions is worthy of summary dismissal. And the implication is that Leibniz's charge that any non-mechanical explanation constitutes a miracle should be seen as nothing more than a cheap rhetorical ploy. Instead, the proper approach to the least particles of matter allows that their hardness is a function of the power

[54] *NPW* 167. Newton does stress that he holds this view "by way of enquiry" which might be taken as support for the "hypothetical" readings of the creation account in *DG*. But it is worth emphasizing the context here, in which Newton is working very hard to differentiate—as he would describe things—his own empirical, evidence-based approach to natural philosophy with Leibniz's dogmatic approach which insists on unverified hypotheses.

of the creator. This seems to reflect the same mindset and general approach as *De gravitatione*.

A fuller version of a very similar dialectic plays out in Newton's draft letter to the editor of the *Memoirs of Literature*. Here he is responding to Leibniz's charge that gravity is an occult quality. Newton's reply focuses on Leibniz's presupposition that the only satisfactory explanation would be a mechanical one. He argues that this presupposition is unwarranted and, as an illustration, points to other features of bodies that he takes to be inexplicable on mechanical grounds. Specifically, he mentions inertia, extension, duration, mobility, and the hardness of fundamental particles (i.e. their solidity or impenetrability). Here is the key passage that follows:

> [Y]et no man ever attempted to explain these qualities mechanically, or took them for miracles or supernatural things or fictions or occult qualities. They are the natural, real, reasonable, manifest qualities of all bodies seated in them by the will of God from the beginning of creation & perfectly incapable of being explained mechanically.[55]

Here the themes of the *De gravitatione* account come through much more clearly. There just is nothing more to solidity or mobility than a power implanted by God in creation. No further explanation is needed, or indeed, possible.

It seems to me that here there is something insightful about Newton's approach to bodies. In one sense, it is commonplace to say that Newton recognized the limits of the mechanical philosophy. This is true just because Newton insisted on gravitation as a force despite the fact that there was little by way of prospects for mechanical explanation of gravitational phenomena. But in passages like those above—and in the *De gravitatione* account more generally—we see a deeper sense in which Newton recognized the limits of the mechanical philosophy.

Newton recognized that there were certain fundamental posits of mechanism that mechanism itself did not explain. Explaining phenomena through appeal to the motions of solid material corpuscles is well and good, but there will still be questions about the solidity and mobility of those corpuscles. And, at some stage, explanations will run out. Some things will need to be accepted as primitives. Consider what Locke says about the solidity of bodies:

> If any one asks me, *What this Solidity is*, I send him to his Senses to inform him: Let him put a Flint, or a Foot-ball between his Hands; and then endeavour to join them, and he will know. If he thinks this not a sufficient Explication of Solidity, what it is, and wherein it consists; I promise to tell him what it is, and

[55] CUL Ms. Add 3968, 257v. Printed at *NCS* #918 and *NPW* 151–152. Underlining added.

wherein it consists, when he tells me what thinking is, or wherein it consists; or explain to me, what Extension or Motion is.[56]

On the one hand, Newton's account seems to channel a similar line of thought. These features of bodies are basic and fundamental. The strength of Newton's approach—to my mind—is that it balances this sense in which certain features of bodies are primitive with an attempt to explain them insofar and only insofar as they can be explained by enumerating them and appealing to the power of God.[57]

4.9. The Account of Bodies in *De gravitatione* as a Theory

As we have seen above, Newton's views on this subject developed over a long period of time. And while there are notable changes, they show a marked consistency from the time of *De gravitatione* through the final edition of the *Opticks*. Part of the reason for exploring texts that are, in some sense, ancillary to *De gravitatione* has been to suggest that we should take Newton's views surrounding bodies in *De gravitatione* very seriously. Put differently, I have been arguing that we should read Newton as being committed to a particular account of the nature of bodies. That being said, my claim is that Newton's career-long commitment is to the broad contours of the view outlined in *De gravitatione* and not to the exact details of the view in that particular manuscript. Put differently, I think we should be willing to discount the hesitance that Newton shows before giving the theory, or that we should read it as hesitance about the details of the general theory offered in that text, rather than as hesitance about the basic structure of the theory itself.

This is reflected in the language I have used to discuss the theory. Recall from above that the account of bodies in *De gravitatione* is commonly referred to as a hypothesis. Using the word "hypothesis" in a context relating to Newton, even if one uses it with the explicit intention of picking out its modern-day meaning, seems to me to risk the introduction of a certain bias.[58] I prefer to talk about the account in *De gravitatione* as a theory. Theories can be somewhat general in their content. They can offer a broad outline which admits of further specification, or of different possible further specifications. And theories are eminently revisable. All told, theories can display a high degree of flexibility. But they are more than hypotheses; one can be committed to a theory and justifiably committed to that theory.

[56] Locke 1975, pages 126–127.

[57] For further discussion of the issues discussed in this section see Connolly 2024. See also Joy 2012. Also helpful is the discussion of Maupertuis' views on these issues in Downing 2012.

[58] For Newton's various uses of the word "hypothesis," see Cohen 1966. See also the useful discussion of hypotheses in Koyré 1965, ch. 2.

4.10. Conclusion

The plan for the next two chapters is to make good on the idea that Newton aimed to outline God's basic strategy for creation in *De gravitatione* but that he also allowed for a certain level of flexibility in understanding precisely how God might have pursued that strategy. The next chapter will delve into the details of the account in *De gravitatione* and offer a reconstruction of the precise view on offer in that text. The following one will look at developments of that theory in later texts and consider Newton's understanding of the relationship between the theory and the work to be achieved in natural philosophy.

5

Bodily Dependence and Bodily Independence

5.1. Introduction

Returning to the question that opened the previous chapter, we can again ask: What are Newtonian bodies? As we saw in Chapter 2, part of the answer to this question involves bodies being secondary substances. So a fundamental component of Newton's answer to this question is that bodies are things which depend on God for their existence. In one sense, this is unremarkable. Nearly every philosopher in early modern Europe would have conceded as much. In another sense, this point is absolutely critical. Newton took the dependence of bodies on God to be incompatible with their possession of an independent substance or essence. This led him to articulate an alternative account of the nature of bodies, one that is highly distinctive and innovative. In this sense, understanding the way in which Newtonian bodies depend on God is essential to understanding what they are.

It is clear as early as *CPQ* that Newton had begun to think about the ways in which dependence on God might be understood as a defining feature of created substances. One of his arguments against the idea that the "first matter" might be an undifferentiated block was that such a block would depend on its parts. But this would fly in the face of the concept of first matter insofar as it would render the "first" matter dependent on something else. Here is Newton's way of putting the point: "For it is a contradiction to say the first matter depends on some other subject ~~(except God),~~ since that implies some former matter on which it must depend."[1] What this passage shows is that Newton, at least at first, seemed to think it was clear that even the most fundamental matter should be understood as relying on God. Of course, Newton canceled the "except God." We need not read this, however, as him backtracking on the claim that the first matter might depend on God. Instead, we might see this as him just recognizing that the theological point was largely irrelevant in this context. It was simply a different type of dependence

[1] *CPQ* 339.

Newton's Metaphysics of Substance. Patrick J. Connolly, Oxford University Press. © Patrick J. Connolly 2025.
DOI: 10.1093/9780191959240.003.0005

that was at issue in his discussion. Whatever the reason for the cancel, the entry shows that these issues were on Newton's mind from a very young age.[2]

Beyond this, however, things become much more difficult. While it is clear that bodies depend on God, the fact is that there are wide variety of forms this dependence might take.

5.2. Degrees of Dependence

The following is a taxonomy of forms of dependence that bodies might have on God. The forms of dependence listed are not exhaustive. Degrees of dependence intermediate between the ones listed are no doubt possible and many of the degrees of dependence listed admit of specification into distinct subspecies.[3] Nonetheless, the list below provides a helpful starting point or frame of reference for thinking about the relevant issues and for situating interpretations of Newton against one another.

[A] **Complete Independence:** Bodies are not dependent on God in any way. They are not created by God, not subject to causal interference from God, and not subject to destruction by God.

[B] **Subject to Causal Engagement:** Bodies are not created by God but can be acted on by God.

[C] **Creationism:** Bodies are created by God but following this initial act of creation God largely ceases to engage with them. This might be understood as a classically Deist view. On this view God is a (mere) *causa secundum fieri* or cause of coming into being.

[D] **Conservation:** Bodies are created by God who must also will that they continue to exist but this exhausts God's causal contribution to the ordinary workings of nature. This view is commonly associated with Durand of Saint-Pourçain (c. 1275–1334).[4] This view makes God a *causa secundum esse* or cause of being.

[E] **Concurrentism:** Bodies are created and conserved by God but also require God to "concur" with their actions in order to produce any effect. In these

[2] Gabbey 2011, page 426, commenting on this passage remarks that the cancellation is a "reminder that that for Newton creation was literally *ex nihilo*." Gabbey is certainly right that Newtonian creation is *ex nihilo*, but I struggle to see how this is affirmed by the cancellation.

[3] For a detailed exploration of this logical space, the comparative merits and demerits of the various positions, and the approaches of various medieval and early modern authors to these issues see Freddoso 1988, Freddoso 1991, and Freddoso 1994.

[4] For a sympathetic analysis of Durand's views on this issue see Tuttle 2022. Also, while Durand is perhaps the best-known medieval proponent of mere conservationism, it is worth noting that he was not alone in endorsing the view. See Frost 2014.

instances of concurrence, both God and creatures make a genuine causal contribution to the production of some effect.

[F] **Quasi-Occasionalism:** Bodies are created and conserved by God. Some of their observed behaviors are brought about by causal powers which they possess and exercise (perhaps with God's concurrence). Other observed behaviors are not produced by bodies but are instead brought about directly by God. Quasi-occasionalism of this sort has recently been attributed to a number of early Newtonians. These figures have been read as holding that bodies themselves are responsible for their "mechanical" behavior but that gravitational forces are produced by God and not by bodies.[5]

[G] **Full Occasionalism:** Bodies are created and conserved by God. But bodies are also entirely passive and possess no genuine causal powers. God is responsible for all activity; the most that can be said for bodies is that they provide the "occasion" for God to reposition them in a certain way. This view is commonly associated with Malebranche.[6] But there were also proponents of the view in Cambridge during Newton's time and it made some inroads among early Newtonians.[7]

[H] **Pantheism:** God and bodies are not properly distinct. Bodies are "in" God or are "part" of God. Their actions are God's actions.[8]

Where does Newton fall on this spectrum of total independence to total dependence? As should already be clear, he certainly rejects views on the far independence side of the spectrum. [A] and [B], for example, are non-starters. They clearly fail on Biblical grounds. Newton took quite seriously the account of creation in Genesis.[9] And my view is that *De gravitatione* should be read as an effort to understand what divine creation of bodies might have entailed.

One might be tempted, then, to think that the range of possible interpretations runs from [C] to [H]. Leibniz, for example, seems to have believed Newton's position was something like [C].[10] But this interpretation is untenable. Of course, Leibniz did not have access to *De gravitatione*. If he had, he would have seen good

[5] Downing 2012, pages 285–286, ascribes this view to Richard Bentley, Samuel Clarke, and William Whiston. See Connolly 2017 for a more detailed treatment of this position in Bentley and Sangiacomo 2018, Wolf 2019, Henry 2020a, and Levitin 2022, pages 745–748, for more detailed treatments of it in Clarke. See also Connolly 2021, page 146.

[6] For overviews of Malebranche's occasionalism see Nadler 2000 and Cunning 2008.

[7] See Turner 1685 and Ablondi 2013, who ascribes the view to Andrew Baxter.

[8] Pantheism, the view that nature is to be identified with God, should be distinguished from panentheism, the view that God is immanently present to all things. Newton's commitment to the latter view is uncontroversial. See Henry and McGuire 2018. For a helpful taxonomy of versions of pantheism and panentheism, see Melamed 2018 or Melamed *manuscript*.

[9] See *CPQ* 447–449 and *NCS* #244 and #247. See also Snobelen 2008 and Janiak 2012. More generally, see Mandelbrote 1994 and Mandelbrote 2006b. See also CUL Ms. Add 3965, 547r: "Matter is not eternal, but had its origin from the will of God."

[10] As did, it seems, Bertrand Russell. See Russell 1945, page 537.

evidence that Newton endorsed some form of conservationism in passages like the following:

> God does not sustain his creatures any less than they sustain their accidents, so that created substance, whether you consider its degree of dependence or its degree of reality, is of an intermediate nature between God and accident.[11]

> That for the existence of these beings it is not necessary that we suppose some unintelligible substance to exist . . . extension and an act of the divine will are enough. Extension takes the place of the substantial subject in which the form of the body is conserved by the divine will.[12]

Similarly, in his "Exposition of 2 Kings 17: 15–16" Newton tells us that God should "be celebrated not so much for his essence as for his actions, the creating preserving & governing all things according to his good will & pleasure."[13] It is not particularly surprising to see Newton affirming conservationism. One would be hard pressed to find an early modern European philosopher who denied it. So, Newton falls somewhere in the range of [D] to [H].

5.3. The Issue in the Secondary Literature

Where does Newton fall on the scale from [D] to [H]? This is a difficult question and opinion is divided among commentators. But there is a marked tendency to attribute to Newton more, rather than less, divine involvement. Some discussion of specific authors will help to make this clear. Howard Stein, Hylarie Kochiras, and Lisa Downing all adopt a form of [F].[14] They believe bodies consist of a set of powers that have genuine causal efficacy but that the powers only remain unified together as a single entity as a result of God's continuous and direct activity. Authors like Martin Tamny, Jonathan Bennett and Peter Remnant, Roger Woolhouse, and Christopher Conn adopt a position closer to [G]. For them, Newtonian bodies cannot even move without direct divine intervention.[15] Liam

[11] *DG* 47.

[12] *DG* 43.

[13] Yahuda Ms. 21, 2r. Underlining added. These notes are related to a sermon, or series of two sermons, Newton may have been required by statute to preach as a master of arts, possibly in Trinity College chapel. See Westfall 1980, page 354 and Levitin and Mandelbrote 2019, pages 341–342. Though see also Feingold 2021, page 183. Levitin and Mandelbrote offer further analysis of the notes as does Iliffe 2017, page 140.

[14] Stein 2002, page 279; Kochiras 2013, pages 331–336; Downing 2014, pages 112–114. Dobbs 1991, ch. 6, claims that Newton endorsed a different form of [F] but only during a circumscribed timeframe. And Hall 1998, page 52, suggests that Newton was highly sympathetic to a different form of [F], but stops short of claiming Newton endorsed it.

[15] Bennett and Remnant 1978, pages 7 and 9; Tamny 1979, page 56, Woolhouse 1982, pages 88–89; Conn 1999, pages 439–440. See also Hall and Hall 1962, pages 79–81; Jalobeanu 2007; Joy 2012,

Dempsey agrees with these authors that bodies require God's continuous and direct activity in order to move and exhibit their strictly corporeal activity, but he nonetheless maintains they have a genuine casual ability to interact with minds. He therefore shows some sympathy with [G] but ultimately offers an alternative route to [F].[16] And Geoffrey Gorham has endorsed a position like [H], allowing that Newtonian bodies can be understood as modes of God.[17] As this brief survey has shown, contemporary readers of Newton tend to understand Newtonian bodies as exhibiting a very high level of dependence on God. He is usually read as holding [F], [G], or [H].[18]

My view is that this dominant strand of thought in the literature is misguided. Newton was very often at pains to stress the tight connection between God and Creation. But this connection is properly interpreted as one of *presence* rather than of *causal influence*. God's immanence was of vital importance to Newton. But this immanence was consistent with bodies enjoying a high level of autonomy and it was consistent with bodies possessing genuine causal powers. As the General Scholium puts the point: "In him are all things contained and move, but he does not act on them nor they on him."[19] In context, the question is about whether bodies experience resistance as a result of moving "in" God, akin to the resistance they would experience moving in a fluid. But zooming out, we can see that Newton's resolutely negative answer to this question points toward a default position of causal non-interference. So, my view is that Newton was committed to [D] and to none of the stronger views outlined above.

My goal below will be to argue for this interpretation of Newton through a critical engagement with positions that attribute higher levels of divine involvement to him. Importantly, issues of divine dependence will not be the sole focus of the following. Because engaging issues of divine dependence requires tackling a number of the key issues in *De gravitatione*, the following discussion will also offer a characterization of the powers that define bodies in that text. In this sense, critical engagement with other positions will provide the basis for my positive interpretation of Newtonian bodies.

page 337; Reid 2012, page 128; Reid 2022, pages 1499–1501. See de Hatzfeld 1724, pages 18–19, for an eighteenth-century interpretation of Newton along these lines.

[16] Dempsey 2014 and Dempsey 2017.

[17] Gorham 2011a, especially pages 308–309, and Gorham 2011b.

[18] Some other authors, whose precise positions are slightly less clear, but who nonetheless conform to this pattern include Snow 1926, ch. 4; Heimann and McGuire 1971; Westfall 1971, pages 396–398; Heimann 1973a; Heimann 1973b; Heimann 1978; McMullin 1978, page 76; Wilde 1982, pages 101 and 107; Kerszberg 1986, page 104; Westfall 1986, pages 233–234; Davis 1991, pages 11–13; Snobelen 2001, page 202; McGuire and Slowik 2012, page 296; Slowik 2013, page 430.

[19] *Principia* 941.

5.4. Rejecting Pantheism

The most extreme view in the literature is that held by Gorham.[20] He arrives at [H], what I have called a pantheist position, as a result of two interpretative moves. The first is an understanding of absolute space as a divine attribute. This ensures that space is not ontologically distinct from God. The second is an identification of bodies with space. This ensures that bodies are not ontologically distinct from space and, transitively, ensures that they are not ontologically distinct from God. While Gorham wants to claim that his view still allows for a "robust distinction" between God and the world, he nonetheless accepts the pantheist implications of his view: "If bodies are quantities of space over time, contingently endowed with mobility, impenetrability, and the power to affect our senses, and space and time are divine attributes, then Newton seems committed to the view that bodies are modes of God. That is, in fact, Newton's view."[21] The radical nature of this interpretation can be emphasized by pointing out its close structural similarity to Spinoza's views. Like Spinoza, Newton will hold that God is a substance, that extension is one of the attributes of that substance, and that bodies are modes of that divine attribute.[22]

We have already, in Chapter 3, seen arguments against the claim that space is a divine attribute. To the extent those arguments are successful, they establish some ontological distance between God and space and, therefore, move Newton away from [H]. But suppose Gorham were right that space is a divine attribute. To the extent that bodies are modes of space, Gorham's position faces serious problems when it comes to the issue of divine immutability. Bodies move. On Gorham's analysis, this means that at some times a given region of space will be, or will instantiate, a body and at other times it will not. This implies that regions of space change over time. To the extent that space is a divine attribute, this suggests that the divine attribute will change over time. Even worse, it will be in constant flux. To the extent that this is correct, the view is incompatible with divine immutability. This is a seriously problematic aspect of the interpretation.[23]

Ultimately, Gorham may be able to avoid this implication of his view.[24] The reason is that he jettisons mobility and impenetrability as genuine powers of bodies. In doing so, he makes bodies nothing more than "spatial occasions" for God to produce various ideas in human minds.[25] So divine immutability is saved and a quasi-Spinozistic view is sidestepped. But only at the cost of a, perhaps rather

[20] Gorham 2011a and Gorham 2011b.

[21] Gorham 2011a, page 308. See also Gorham 2011b, page 27.

[22] Of course, Gorham's Newton will still be non-Spinozistic in many other ways. To pick just two notable differences, Gorham's Newton will reject a material plenum and necessitarianism.

[23] Leibniz forwards a version of this argument against Clarke's claim that space is a divine property. See Leibniz and Clarke 2000, page 45.

[24] I owe this point to Kochiras 2013, note 28.

[25] Gorham 2011b, page 24.

Berkeleyan, idealism.[26] The details of Gorham's claim that mobility and impenetrability are superfluous and play no genuine role in Newton's account of body will be critically assessed below. But, for the moment, it is perhaps just worth emphasizing what a radical position this is.

In part, the view is radical because it identifies space with God; or as Gorham would say, it assimilates space to God. As argued in Chapter 3, this part of the view should be rejected. But the view is also radical because it identifies bodies with parts of space. Or, as Gorham might say, it assimilates them to parts of space. The next section shows that many other commentators make a similar identification and argues that we have reason to reject this part of the view as well.

5.5. Empowered Spaces vs. Spatially Situated Powers

Our starting question in this section is about the relationship between the powers that constitute Newtonian bodies and space (or, perhaps, places in space). Some have suggested that specific regions of space are themselves endowed with powers. This view establishes a very tight connection between certain spaces and "their" powers; a specific region of space instantiates the power.

Bennett and Remnant, for example, have it that the account of bodies in *De gravitatione* begins with God endowing regions of space with impenetrability.[27] Similar language is employed by Martin Tamny, Alan Gabbey, Geoffrey Gorham, and Rob Iliffe.[28] And an account along these lines is found in Christopher Conn, where God creates by "conferring" impenetrability on a region of space.[29] In Howard Stein, the process of creation involves making a region of space itself impervious to other bodies.[30] The issue is less clear in other authors. But many might be read as endorsing something like this view.[31]

If this is Newton's position, then it runs into an immediate problem. Bodies can move. But Newton is adamant that parts of space cannot move. Parts of space are defined by their position. If a part of space were, *per impossibile*, to move it would cease to be that part of space. It would lose its identity.[32] All of this coheres neatly with Newton's view that motion is transference from one part of space to another.

[26] As Gorham 2011b, page 30, describes it, a "quasi-idealistic ontology of matter." This points to similarities between Gorham's view and that of Dempsey in Dempsey 2014. Robles 2000 and Robles 2001 also offer a Berkeleyan reading of *DG*.

[27] Bennett and Remnant 1978, page 8.

[28] See Tamny 1979, page 56; Gabbey 2001, page 7; Gabbey 2011, page 435; Gabbey 2016, pages 444–445 and also 442 where God's action is that of "'solidifying' space"; Gorham 2011a, page 293; Iliffe 2017, pages 103–104.

[29] Conn 1999, page 438.

[30] Stein 1970, page 275, see also page 276 and Stein 2002, page 278 where the qualities constituting bodies have regions of space as the metaphysical support of their being.

[31] See, e.g., Joy 2012, pages 327–328.

[32] See Newton's discussion of the issue at *DG* 39. See also Nerlich 2005.

But it does not cohere with the idea that a body is created by endowing a part of space with impenetrability. Insofar as the part of space is incapable of moving, the impenetrability will also be incapable of moving. And the resulting object, insofar as it is immobile, will fail to be like a body.

The problem just outlined is well-known and has led a number of commentators to converge on a solution. Rather than allowing that a part of space moves, God might first cause one part of space to cease being impenetrable and then immediately cause another adjacent space to become impenetrable. Thus the "body" would appear to move. One key upshot of this approach is that bodies will be hugely dependent on God. They will need to be recreated by God at every moment, or at least at every moment in which they are in motion. Stein, for example, writes that in order to create mobility in a body God must:

> Having established a region of impenetrability, allow it, in the course of time, to migrate from one place to another; or more precisely, confer the property of impenetrability, not on a fixed spatial region, but on different regions at different times—and in such a way that the mutations of the distribution of impenetrability constitute a smooth motion in space.[33]

As Tamny, who provides a similar analysis, says: "God is then very active indeed."[34] The universe becomes almost cinematographic. Just as individual still pictures when projected in quick succession create a "moving image," the singular states of the universe, while in fact disconnected from one another, occur in succession so as to create apparent continuity and motion.[35]

One way to further bring out the implications of this reading is to note Christopher Conn's argument that the account of bodies offered in *De gravitatione* entails four-dimensionalism about bodies.[36] Three-dimensionalists hold that bodies *endure* in the sense that they exist in their entirety through the length of time in which they exist. Four-dimensionalists hold that bodies *perdure* in the sense that they have temporal parts extended through time which unite to form the body. If it is true that bodies are distinct regions of space successively endowed by God with impenetrability then Conn's argument is highly plausible. It is the only way to reconcile the following two claims: (1) a body at a given time is a region of space endowed with impenetrability; (2) one and the same body can exist at numerically distinct places. We will return to this implication of the interpretation

[33] Stein 1970, page 275.

[34] Tamny 1979, page 56.

[35] The language of a cinematographic universe here is borrowed from an analysis of Descartes' views by Martial Gueroult, who in turn borrowed the phrase from Bergson. See Gueroult 1984, volume 1, page 194.

[36] See Conn 1999, especially pages 441–442. Conn can profitably be read as supplying an answer to the problem identified at Tamny 1979, pages 56–58.

under consideration below. But for the present moment, it is worth just emphasizing again that this account entails that the motion of bodies depends on God in a fundamental way; it is nothing other than God's subsequent creation and elimination of each of their spatio-temporal parts.

Seen in this way, one motivation for reading Newton as an occasionalist stems from a certain understanding of how the powers of bodies relate to space. Insofar as this is correct, an alternative understanding of the relationship between the powers and space has the potential to remove this motivation and afford bodies a higher degree of independence.

As it happens, just such an alternative understanding is available and it provides a superior interpretation on both textual and philosophical grounds. The approach just considered had it that there were certain things, regions of space, which were endowed with powers. The upshot of this was to foreground the regions of space in the account. Iliffe, for example, has it that the *spaces themselves* are the bodies.[37] And Conn claims that it follows from this account that "an atom *is* a region of space."[38] Call this the *Empowered Spaces* view. On the alternative approach that will be defended here, the powers are not powers *of* regions of space. Rather, they are powers *located in* regions of space. The emphasis here is on the powers themselves making their relation to space secondary. Call this the *Spatially Located Powers* view.

To be fair, the initial presentation of the theory of bodies in *De gravitatione* does leave something to be desired and can lend support to an *Empowered Spaces* interpretation. But the fact is that, in a number of places, Newton works carefully to clarify his meaning, to specifically guard against this interpretation, and to signal that he favors the *Spatially Situated Powers* view. Understanding this requires a brief tour through the initial treatment of body in *De gravitatione*.

The account begins, on my interpretation, with a series of intuition pumps. Newton first asks us to think about our own volitional agency and the changes it produces in the world.[39] He then notes that whatever agency we have must be had by God and be had by God to an even greater degree. When we move our bodies into empty spaces, we thereby prevent other bodies from occupying those spaces. God, of course, does not have a body but should nonetheless be capable of producing the same effect—making a region of space impervious to bodies—directly. Newton next notes that divine activity of this sort would produce something with many features had by bodies. To this point, some of Newton's characterizations do make it sound like the bodies in questions are spaces. Newton writes about God

[37] Iliffe 2017, page 104: "A world made up of these sorts of "endowed spaces" would be indistinguishable from the world in which we now live because such spaces would be able to 'excite various perceptions of the senses and the fancy in created minds, and conversely be moved by them.' These spaces would either be bodies or be like bodies."

[38] Conn 1999, page 438. Emphasis in original.

[39] *DG* 42.

causing some space to be impervious to bodies. And he says that we would consider this space to be a body. I suspect that proponents of the *Empowered Spaces* view have been deeply influenced by these initial characterizations.

But even in the next paragraph we can begin to notice an important shift. Newton points out that the sort of thing described up to this point would not really be a body. The reason is that it would be immobile. What would be needed to get a body would be to "suppose that that impenetrability is not always maintained in the same region of space."[40] Here the thing that is the body is no longer the space, but rather the impenetrability itself.

In the next paragraph, Newton briefly switches back to talking about spaces being the things that constitute the bodies. But he then goes on to offer what it is clearly meant to be his definitive characterization of bodies. He writes that bodies can be defined not as spaces with certain features, but rather as *quantities of extension* with certain features. What comes next contains an explicit articulation of the difference between these two and why it is important. The first of the features had by these quantities of extension is "that they be mobile, and therefore I did not say that they are numerical parts of space which are absolutely immobile, but only definite quantities which may be transferred from space to space."[41] We ought to take this formal definition of bodies seriously. And we ought to take this clarification from Newton quite seriously as well. He specifically warns us against identifying bodies with spaces. Thus, I see this important passage as supporting the *Spatially Situated Powers* view over the *Empowered Spaces* one.

Other portions of text support this interpretation. One instance comes after Newton's first full characterization of bodies as being impenetrable, mobile, and capable of interacting with human minds. Here he describes them as "nothing other than the effect of the divine mind produced in a definite quantity of space."[42] Here, the body is something produced *in* a quantity of space, rather than something pertaining directly to a part of space itself. At a later stage of *De gravitatione*, he reminds his readers that God generates new bodies "in" rather than "from" or "out of" space and that "spaces are not the very bodies themselves, but are only the places in which the bodies exist and move."[43] Again, the distinction is clear. Katherine Brading has put the point beautifully: "As Newton himself says, body is that which *fills* place; a portion of matter, or a body, is not identified as an impenetrable region of space."[44]

[40] *DG* 42.

[41] *DG* 43. Intriguingly, in the manuscript "*definitas quantitates*" was inserted where Newton had first written, then canceled out, "*partes numericas*." This may have been a mere momentary lapse of concentration. But it might also be the case that Newton himself was struggling to articulate the point and move from his initial characterization to the precise formulation of the view.

[42] *DG* 42.

[43] *DG* 50.

[44] Brading 2012, page 25.

A further passage helpful for understanding this issue comes when Newton uses a comparison to the Scholastic framework which would have been familiar to his contemporaries as a way to elucidate his views. Here is the relevant passage:

> Between extension and its impressed form there is almost the same analogy that the Aristotelians posit between prime matter and substantial forms, namely where they say the same matter is capable of assuming all forms, and borrows the denomination of numerical body from its form. For so I posit that any form may be transferred through any space, and everywhere denote the same body.[45]

Newton is quick to point out some of the shortcomings of the analogy, particularly the fact that space has "more reality" than prime matter and can be understood almost as a "form" itself.[46] But, however imperfect, the comparison is still revealing.

One reason is that it breaks the connection between the powers (the form) and any one particular region of space (the matter). Bodies do need to be extended somewhere in space, but they are, as it were, indifferent to any one particular region of space. Just as a substantial form can combine with any prime matter to create an Aristotelian substance, Newtonian powers located at any region of space will be a body. Second, it is the powers (the form) that determine where and what a body is. The "denomination of numerical body" comes from them, not from any particular space. And third, one and the same set of powers (form) can be translated or transferred through space while maintaining its identity. Thus, I take it that Newton's attempt to compare and contrast his view with hylomorphism is more favorable to the *Spatially Located Powers* interpretation than to the *Empowered Spaces* one.[47]

Still further considerations can be drawn from arguments offered by Kochiras.[48] She points out that Newton offers characterizations of space that are at odds with parts of space being or becoming bodies. First, Newton includes among the "principal truths of metaphysics" that God "has created bodies in empty space out of nothing."[49] But if the parts of space were themselves endowed with powers, God would not have made bodies "in" space but rather partially "out of" space and not "out of nothing." Second, space's essential features include its immutability and inability to produce effects. Recall that in the previous section Gorham's interpretation was criticized because it threatened divine immutability. If space were a divine attribute, and bodies were modes of space, then the fact that the modes of space would shift and modify over time as bodies moved entailed that there would

[45] *DG* 43–44.
[46] *DG* 44.
[47] For a complementary, though slightly different, take on these issues, see Hill 2003, sec. 2. To my mind, Hill overstates the extent to which Newton sees himself as endorsing a version of Scholastic hylomorphism. But his analysis is nonetheless illuminating.
[48] See Kochiras 2013, page 332.
[49] *DG* 45.

be change in the divine being. Here the problem is analogous. Newton is very clear that space is immutable in its very nature.[50] But if one part of space, at a certain time, is or becomes a body and, at another time, ceases to be or to constitute a body, then it is very hard to see how space is not itself undergoing change. The *Empowered Spaces* view is therefore difficult to square with the clear characterization of space in Newton's texts. By contrast, the *Spatially Located Powers* view does not face this difficulty. On this view, the bodies are spatially-extended powers with determinate quantities; they are not themselves spaces or places. And while they do occupy different regions of space at different times, their doing so does not result in any substantial change to space itself. Put differently, on the *Spatially Situated Powers* view, spaces undergo mere Cambridge changes as a result of bodily motions.

Recall from above that the *Empowered Spaces* interpretation is committed to four-dimensionalism about Newtonian bodies. It is worth noting that the *Spatially Situated Powers* reading has no such commitment. While I am confident Newton would endorse the *Spatially Situated Powers* reading over the *Empowered Spaces* reading, I think his response to the issue of four-dimensionalism about bodies is less clear. There are, however, some texts which bear on the issue and which motivate a three-dimensionalist interpretation.

We can first look at two passages which bear directly on persistence through time. In a passage added to the General Scholium in the third edition of the *Principia*, Newton is explicit that three-dimensionalism is the correct understanding of the way in which God and human souls persist through time. Here is the relevant text:

> Every sentient soul, at different times and in different organs of senses and motions, is the same indivisible person. There are parts that are successive in duration and coexistent in space, but neither of these exist in the person of man or in his thinking principle, and much less in the thinking substance of God. Every man, insofar as he is a thing that has senses, is one and the same man throughout his lifetime in each and every organ of his senses. God is one and the same God always and everywhere.[51]

The passage is slightly more opaque than one might like. And Newton does, somewhat confusingly, allow that there are spatial and temporal parts.[52] But the clear message is that neither spatial nor temporal parts are properly considered to be constitutive of (to "exist in") the truly unified and indivisible things which persist

[50] *DG* 41.

[51] *Principia* 941.

[52] An analysis of the sense in which Newtonian secondary substances have spatial parts is offered in Chapter 7.

through time. Of course, the mere fact that God and human souls endure, rather than perdure, cannot decisively settle the issue regarding bodies. There are differences in the ontic status of these three types of substance and their method of persistence through time could potentially be one of them. But evidence from *Tempus et Locus* suggests it is not. In that text, Newton makes claims very similar to those in the General Scholium but generalized to include all beings:

> For the Duration of a thing is not its flow, or any change, but permanence and immutability in flowing time. All things endure in so far as they remain the same at any time. <The duration of each thing flows, but> its <enduring> substance does not flow, and is not changed with respect to before and after, but always remains the same.[53]

This passage is even more clearly three-dimensionalist in its orientation. And the fact that it occurs in a context where Newton is emphasizing that temporality and spatiality are common affections of all beings clearly entails that bodies fall under the scope of its claims.

We should also consider an implication of the *Empowered Spaces* view rarely addressed by its proponents. Specifically, if the motion of bodies is just God making different portions of space impenetrable at different times, then why is it that bodies always move through continuous regions of space? Put differently, what, on this interpretation, would be the problem with a body moving by instantly "jumping" from one side of a room to the other without passing through any interceding spaces? Proponents of the view might claim that this is simply not the way in which God has decided to move bodies. Or they might argue that we do not observe any bodies behaving in this way. These are both fair points. But, especially on a four-dimensionalist reading, it is hard to see what sort of *in-principle* obstacle there might be to this sort of movement.

There is a Newtonian text which touches on this issue. It comes from notes on one of Newton's favorite topics: the errors of Catholicism.[54] The relevant section of the text discusses the implausibility of transubstantiation, in particular the fact that it requires belief in ten distinct miracles for each act of consecration.[55] Here is one of the miracles: "I may ad to make christs body translated in instanti wthout passing

[53] McGuire 1978b, page 117.

[54] On Newton's anti-Catholicism, see Iliffe 1999.

[55] As it happens Newton only ends up enumerating, by my count, six miracles. It is worth noting that anti-Catholic themes, particularly critiques of transubstantiation, were common among the divinity acts overseen by Joseph Beaumont during Newton's time in Cambridge. Newton himself was Respondent for a divinity act under Beaumont that focused on transubstantiation, although seemingly not on these metaphysical issues. See Levitin and Mandelbrote 2019. Other places in which Newton is critical of transubstantiation include Yahuda Ms. 15.3, 47v; Yahuda Ms. 15.5, 97r; and Yahuda Ms. 9.2, 99r–99v. Newton also dog-eared a page in his copy of Locke's *Essay* which was critical of the doctrine. See Locke 1690, page 357. *NL* #967.

through intermediate places or removing from its own place, or els to be new produced i.e. created."[56] One could read this as Newton claiming that if an object (in this instance, Christ's body) were to move from one place to another without passing through the intermediate places this would involve something *different in kind* from ordinary motion. As Newton defines motion in the "Lawes of Motion," written sometime in the mid-1660s or early 1670s, for bodies "their passing out of one place or part of space into another, through all the intermediate space is their motion."[57] Transubstantiation would require God to do something different, not to move bodies, but to *recreate* or *newly produce* them in the new space. But, on the *Empowered Spaces* view, this is precisely what happens in ordinary cases of motion, God recreates or newly produces impenetrability in a new region of space. So the contrast Newton wants to draw would be absent.

The two considerations adduced against four-dimensionalism here, some of Newton's comments about persistence and some claims about motion extracted from Newton's views on transubstantiation, are less than fully conclusive. Nonetheless, I do think they constitute admittedly defeasible, but nonetheless strong, evidence that Newton inclined toward a three-dimensionalist approach to bodies. Insofar as this is correct, these considerations constitute another argument against the *Empowered Spaces* interpretation. At a minimum, it is incumbent on proponents of the *Empowered Spaces* interpretation to offer readings of the texts considered here that make them consistent with their view.[58]

5.6. Mobility as a Power

It was noted above that a certain motivation for occasionalism might come from the view that impenetrability is a property *of* a part of space, rather than a power located in space. But one might hold the preferable interpretation, on which impenetrability is merely a power *located* in space, and still use considerations relating to motion to infer that Newton must be committed to a high degree of divine involvement. Consider, for example, the position taken by Lisa Downing:

> God can . . . prevent bodies from entering a certain space, defined by certain limits. That space is now impenetrable, Newton suggests. Further, God could transfer the area of impenetrability according to certain laws. Note that this is not to transfer the portion of space; Newton emphasizes that the parts of space are

[56] Yahuda Ms. 14, 20r.
[57] CUL Ms. Add 3958.5, 81r. Underlining added. Printed at *NCS* #349.
[58] Alternatively, they might concede that these texts are inconsistent with their interpretation of *DG* and instead claim that Newton's views underwent a substantial shift in the years after he wrote *DG*.

immoveable. What is transferred (through different parts of space) is rather the impenetrability.[59]

Downing correctly emphasizes that impenetrability is a power located in space, rather than a power belonging to any particular region of space. And aspects of this characterization line up neatly with the passage from *De gravitatione* that Downing cites here: "[W]e should suppose that that impenetrability is not always maintained in the same part of space but can be transferred here and there according to certain laws, yet so that the quantity and shape of that impenetrable space are not changed."[60] But there is one key difference. Whereas *De gravitatione* is silent on the mechanism underlying the acts of transferring, Downing makes God the agent directly responsible for transferring the power of impenetrability between different regions of space. In this respect, Downing's interpretation is still very similar to those considered above.[61]

My view is that mobility is a power of Newtonian bodies on a par with impenetrability (and the power to interact with minds).[62] Recognizing that this is the case can help to dissolve some of the puzzles surrounding the motion of Newtonian bodies and can thereby undercut the need for continuous divine involvement in Newton's account.

In the previous section, I suggested that the initial presentation of the theory of bodies in *De gravitatione* had a certain staging. Newton first offered a series of intuition pumps. These were designed to help readers begin to form some conception of the kind of creative agential power God might have by reflecting on features of their own agency. They were also designed to help readers isolate and attend to certain key features of bodies, ones that are central to our understanding of what bodies are and that might serve to explain various phenomena associated with them. Following this, Newton offered what I characterized as his official definition of bodies, on which they were determined quantities of extension endowed with certain features.

My suggestion above was that too much attention to the initial stages of this presentation and insufficient attention to the latter stages might lead us astray. It might tempt us toward an acceptance of the *Empowered Spaces* view even though Newton meant to endorse the *Spatially Located Powers* view. Here again, I think there is a worry that Newton's order of presentation runs the risk of muddying the waters.

[59] Downing 2014, page 106.

[60] NPW 42.

[61] Dempsey 2014 is another commentator who correctly understands the relations of powers to space but who nonetheless makes God directly and continuously responsible for the motions of bodies. A somewhat more nuanced, but I believe ultimately similar, view is found at McGuire 1982, pages 178–179.

[62] Katherine Brading also articulates a solution along these lines. See, especially, Brading 2012, pages 24–25.

It is true that Newton's initial pass at bodies in *De gravitatione* gives a certain priority to impenetrability. It begins by thinking about bodies as things that prevent other bodies from co-locating with them and only later moves on to consider issues relating to motion and mind-body interaction. One might take this to mean that impenetrability has some special status in the account or that it is ontologically or explanatorily prior to the other features of bodies. But we are not forced to this conclusion. It is worth remembering just how distinctive and revisionary Newton's approach to bodies was. He was proposing something that diverged quite radically from what would have been more familiar conceptions in the period. Seen in this light, Newton may have been working to ease his readers into things, offering certain thought experiments, and introducing the theory piecemeal in order to make it more readily comprehensible.

Following on from this introductory work setting up his approach to bodies, we get Newton offering his most careful and considered presentation of the view. Part of this was quoted above, but it will now be helpful to examine it in greater detail:

> If they are bodies, then we can define bodies as *determined quantities of extension which omnipresent God endows with certain conditions.* These conditions are: (1) that they be mobile, and therefore I did not say that they are numerical parts of space which are absolutely immobile, but only definite quantities which may be transferred from space to space; (2) that two of this kind cannot coincide anywhere, that is, that they may be impenetrable, and hence that oppositions obstruct their mutual motions and they are reflected in accord with certain laws; (3) that they can excite various perceptions of the senses and imagination in created minds, and conversely be moved by them.[63]

This is a rich passage and there is much to say about it. But we can begin just by noting that here Newtonian bodies are defined by three distinct powers: mobility, impenetrability, and mind-interactivity.

The suggestion here and, I believe, elsewhere in *De gravitatione* is that these are distinct powers and that they are all, as it were, on the same level as one another. My hope is that recognizing this, and thereby giving mobility its due, will pay dividends. As we saw in the previous section, the problem of whether or not the motion of bodies requires parts of space themselves to move is headed off by Newton's more careful formulation. The characterization of mobility here makes it explicit that bodies are not spaces, but only definite extended quantities of powers which themselves are subject to movement.

Looking at some of Walter Charleton's views can help us to get a sense for mobility as an autonomous power of bodies instilled in them by God at creation. In the chapter of his *Physiologia* devoted to "The Motions of Atoms," Charleton spends

[63] *DG* 43.

some time sorting out the wheat from the chaff in Epicurus' teachings regarding motion.[64] Among the views to be rejected are that atoms existed eternally and possess an inherent principle of motion. These should be replaced with the views that "*Atoms were produced* ex nihilo, *or created by God*" and that "*at their Creation, God invigorated or impregnated them with an Internal Energy, or Faculty Motive.*"[65] A similar claim occurs later in the work when Charleton argues that the mobility of compound bodies is a result of "the essential *Mobility* of Atoms."[66] Again, the picture is that a "MOTIVE VIRTUE" or "essential Faculty" of mobility is one of the basic ingredients required for the composition of a material object.[67] My claim is that Newton means to be channeling a similar line of thought.

The interpretation being developed here, on which Newtonian bodies are collections of these three distinct powers located in space, is incompatible with interpretations that make God the immediate cause of the motions of bodies. Suppose we allow, for the sake of argument, that God really is the cause of motions in bodies. We might be able to reconcile this with Newton's claims about mobility in the block quotation above. It only requires that they be mobile and translatable from place to place. But it will be harder to square this with what Newton says about impenetrability. On this view, in what sense is impenetrability a genuine feature of bodies? The worry is that it becomes entirely explanatorily redundant. If God is responsible for all the motions of these spatially extended entities, then the fact that no two of them ever coincide is not due to any fact about *them*. On the contrary, it is just a function of the way in which God has decided to move them. The bodies are not genuinely impenetrable; God simply does not allow any two or more of them to co-locate. Gorham, who like many other commentators is committed to God being directly responsible for the motions of bodies, has noticed a version of this problem. He claims that mobility and impenetrability "do no independent work of their own" and infers from this that Newton's theory actually has no commitment to either of them as genuine or independent powers.[68]

To my mind, this simply cannot be squared with the text of *De gravitatione*. Newton's clearest statement on the metaphysical nature of bodies indicates that they are composed of three powers. Impenetrability, in particular, is throughout the text afforded a great deal of importance as a feature of bodies. I think we should be extremely hesitant to cast it aside as a genuine power of Newtonian bodies given how consistently Newton presents it as something that bodies themselves account

[64] Charleton uses the metaphor of winnowing wheat from chaff with respect to this specific task at Charleton 1652, page 43.
[65] Charleton 1654, page 126.
[66] Charleton 1654, page 269.
[67] Charleton 1654, page 269. See also the similar claims at Charleton 1652, pages 46–47.
[68] Gorham 2011b, pages 23–24. Elliot Chen has noted a related problem. On his view, there is a danger that the laws of motion render the power of impenetrability redundant or unnecessary. See Chen 2020, sec. 2. My view, presented in Chapter 6, is that this misunderstands the priority relation that obtains between powers and laws.

for. In the block quotation above, there are oppositions that obstruct the mutual motions of bodies. Elsewhere, Newton has it that a body can "resist the motions of bodies and perhaps reflect them" and will "be capable of reflecting and being reflected."[69] And in a place where he is working to differentiate his views from those of Descartes, he insists on the power of bodies to stimulate ideas in our minds and warns that "to remove this [power] is no less than to remove that other faculty by which bodies are enabled to transfer mutual actions from one to another, that is, to reduce body into empty space."[70] Later in this stage of the argument, he consistently uses the plural in speaking about the capacities (what I have been calling powers, and what the Latin has as *facultates*) that constitute bodies.[71] On the readings being considered, this impenetrability of bodies will be illusory and this will render Newton's formulations deeply infelicitous.

If this is correct, then we would do well to reject occasionalist readings and accept that bodies are constituted by distinct powers of impenetrability, mobility, and mind-interactivity. Allowing that this is correct will, in turn, solve another problem posed for Newton's account. Bennett and Remnant have argued that Newton is attempting to launch an ontology of bodies with a single relational property.[72] For them, impenetrability is tasked with both (a) managing a body's situation in space (its extension taking a particular shape or form and differentiating empty space from non-empty space) and also (b) ensuring that the body excludes other bodies from that space. They allege that no one thing can accomplish both of these tasks.

I confess that, despite considerable effort, I have never been able to feel the force of this objection. It seems to me that a spatially distributed power of impenetrability is sufficient to the task. Its arrangement in space *fills* a given space, thereby sufficing for the differentiation of the space it occupies from empty space. And its impenetrability guarantees that it will not co-locate with any other similarly spatially distributed power.

Downing gives voice to the problem in the following way:

> [A] spatially distributed counterfactually characterized property instantiation must approach and be deflected. And the instantiated property in question is the disposition to deflect other such instantiated properties. But what is the difference between this property instantiation being deflected and not being deflected? Between the property instantiations overlapping and their not overlapping? There is no such difference. And if there is no such difference, then we cannot make sense of the disposition in the first place.[73]

[69] *DG* 42.
[70] *DG* 49.
[71] *DG* 50.
[72] Bennett and Remnant 1978, pages 7–9.
[73] Downing 2014, page 107.

Again, I struggle. On my understanding, the difference between them deflecting and not-deflecting just is the difference between them succeeding in or failing to occupy the same place. The nature of absolute space and the logic of location in space ensure that there is a difference.

Suppose, however, that there is a problem here. My failure to see it may stem from the fact that I see mobility as a power distinct from impenetrability and from the fact that this suffices to solve the problem.[74] As mentioned above, Newton's order of presentation in *De gravitatione* is slightly misleading in that the initial presentation of the view leads with impenetrability. But note that in Newton's more careful formulation of his view, the one captured in the block quotation above, it is *mobility* that comes first. Mobility is what makes a body a *definite quantity*, one whose size and shape remain stable, that can be transferred from place to place. The next condition, impenetrability, holds that two of *this kind* cannot spatially co-incide. This ensures that bodies are built out of more than one property.

Consider what might happen if God wanted to create a different world, one much simpler and calmer and more peaceful than our own. Rather than making bodies, in this world God opted to make *phasmas*. A phasma consists of nothing more than a power of mobility distributed in a determinate way in space.[75] The regions occupied by phasmas are differentiated from empty space insofar as there is something in them: a power of mobility. Imagine now that two of these phasmas are moving directly toward one another (on what, in our world, we would call a collision course). Because the phasmas are nothing other than the spatially extended power of mobility there is nothing preventing their co-location. So after a moment of brief overlap, in which they occupy the same space, they continue along their respective trajectories. Nothing exists which is sufficient to explain why anything else should happen.

God might have made this world. Though, if God did, we would not know about it. Even if we existed in such a world, the phasmas were not endowed with a power to interact with our minds and "We should not know Bodies for substances, were it not for their sensible qualities."[76] We will be as the proverbial ships in the night. Rather than phasmas, God made bodies. But we can think of bodies as modified phasmas. In addition to a spatially located power of mobility, bodies also have a power of impenetrability. This ensures that their mutual meetings look different

[74] Again, see also Brading 2012, pages 24–25.

[75] Some might worry that there is something incoherent in the notion of a phasma. But there are passages where Newton seems committed to the view that God could make something like this. E.g., at *NCS* #988 he writes that: "One may suppose that God can create a penetrable body & so reject the impenetrability of matter." And at *NCS* #989 he writes that: "One may suppose that there may be bodies penetrable or immoveable or destitute of force, or with attraction mutually unequal." I think we should reconstruct Newton as thinking that bodies are collections of powers created by God, that God had free choice in selecting which powers to create, and that God's omnipotence ensures any non-contradictory collection of powers could be created.

[76] *NCS* #40.

than mutual meetings of phasmas. Seen in this way, the problem should hopefully dissolve.

Gorham and Downing take a different route toward solving the problem. For them, it is solved by the fact that bodies also possess the power of interacting with human minds. They find in this the needed second quality or aspect of bodies.[77] I also recognize the ability to interact with minds as a genuine power of bodies. So, if this is sufficient to solve the problem, then I am happy to endorse this solution. But, as should be clear from above, I also want to argue that this is not the only solution on offer.

I propose the following diagnosis. Interpretations of Newton that make God directly responsible for the motions of bodies, interpretations like those offered by Bennett and Remnant and Downing, will unsurprisingly struggle to see both mobility and impenetrability as genuine and distinct powers of bodies. In one way or another, they will outsource some of the work Newton believed these powers themselves accomplished to God. And this leads them to charge Newton with having an account of body not fit for purpose. The way out of the difficulty, on my interpretation, involves recognizing the autonomous causal powers of bodies.

5.7. A Deflationist Account of Bodies

Recall Kochiras' deflationary account of substance in Newton discussed in Chapter 3. Her theory allowed that Newton endorsed the existence of various types of substances: God, minds, and bodies. And it was deflationist insofar as these substances had no central essence or substratum. I argued that while this was a promising approach to secondary substances, it was the incorrect position regarding God. The aim of this section is to further the claim that this is a promising approach to Newtonian bodies.

The basic appeal for this kind of approach to bodies comes from the fact that Newton offers a very clear rejection of prime matter as a being that underlies secondary substances. And some of his reasons for rejecting prime matter generalize in a way that suggests he would also strongly reject other approaches that posit a "substratum" undergirding substances. This motivates an identification of substances with their powers or with their attributes. As Newton writes, "substantial reality is to be ascribed to these kinds of attributes, which are real and intelligible things in themselves and do not need to be inherent in a subject."[78] Some type of deflationism is clearly called for.

Kochiras' version of deflationism has it that substances are to be identified with their characteristic attributes. Unfortunately, Kochiras never specifies precisely

[77] Gorham 2011b, page 23 and Downing 2014, page 108.
[78] *DG* 47.

what qualifies something as a *characteristic* attribute. On the one hand, something may be a characteristic attribute of a certain type of substance in a very minimal sense. On this view, a characteristic attribute of a substance is just an attribute that substances of that type tend to have. If this is all that is meant, I fully endorse the proposal as it applies to Newtonian bodies.

But the language of characteristic attributes might denote something stronger. Particularly relevant here is the Cartesian background. In *Principles*, Part 1, sec. 53, Descartes claims that each substance has a *principal attribute* and that this "principal property . . . constitutes its nature and essence."[79] My view is that we ought to be very careful of any reading of Newtonian characteristic attributes that makes them akin to Cartesian principal attributes.

Certainly Newton rejected the specifics of this Cartesian position, adamantly denying, for example, that a body could be understood through a primary attribute of extension. And Newton also claims that at least one characteristic attribute, namely being extended in space, is shared between bodies and minds. But, leaving aside the specifics of Descartes' view, it also seems that Newton was not attracted to the general strategy. Part of what makes something a Cartesian principal attribute of body is the fact that all other features of bodies are meant to bear some tight connection to that attribute. For Descartes, the primary attribute of body will be extension. And this means that other features of bodies—their weight and hardness and so on—either reduce to, are explicable in terms of, or are intelligible in terms of extension. And something similar will be true for the substance mind and its principal attribute of thought. Newton rejects this approach. He posits distinct powers of bodies to do distinct kinds of work. And the powers seem importantly heterogenous. That is to say, we cannot understand or explicate all of them with reference to some one primary one.[80] Indeed, *De gravitatione* seems to contain an anti-Cartesian argument specifically to this effect.[81]

On the one hand, it may be fairly straightforward to notice that Newton differs from Descartes in this way. But focusing on the difference does help to bring out an important feature of Newton's account of bodies. Specifically, it focuses our attention on the fact that Newtonian bodies have no natures or essences beyond the divine choice about which powers to locate and combine together. They just are collections of powers. God could have presumably come up with any other number of different powers and could have combined them in any number of ways to generate a veritable farrago of types of secondary substances. But the essence of any given secondary substance will simply be exhausted by the powers that compose it.[82] This point will be particularly important later when we consider

[79] *CSM*, I, 210.
[80] This point is briefly made by Biener and Smeenk 2012, page 114.
[81] *DG* 48. For an alternative approach to this issue, see Gorham 2011b, where it is argued that the ability to cause perceptions in minds is the principal attribute of Newtonian bodies in precisely this Cartesian sense.
[82] For more on this, see Connolly 2024.

the ontology of human beings and the senses in which Newton should or should not be considered a dualist. For now, the point is important because it provides insight into the sense in which Newtonian bodies, unlike their Aristotelian and Cartesian counterparts, do not have robust essences or natures. And this is another lens through which the Newtonian account of bodies can be seen as deflationist.

While Kochiras does not clearly specify what is meant by a characteristic attribute, it seems likely that she has the minimalist, rather than the Cartesian, sense in mind. Insofar as this is the case, the proposal is on firm footing. This then leaves us with the view that Newtonian bodies just are collections of powers. In *De gravitatione*, they are the powers of mobility, impenetrability, and interaction with minds.

5.8. The Glue Problem

Newton's theory, as he himself points out, is radical in that he abandons a traditional conception of a substance underlying the powers or properties of bodies. There are a number of reasons that Newton might have been attracted to the idea of foregoing a commitment to substances or substrata. First, he saw a tension between the idea that bodies could have a free-standing independent entity at their core while also being fully dependent on God. So his understanding of bodies encoded a suitable form of piety and blocked a route to atheistic materialism. Second, the idea of a substance had been the subject of considerable philosophical scrutiny in Britain by the time of *De gravitatione*'s composition.[83] Third, there are obvious considerations stemming from parsimony that motivate a view like Newton's.

This is all well and good. But it is worth recalling that substances had not been introduced as an idle posit. Substances had traditionally been thought to do important philosophical work.[84] One traditional job of substances was to play a *unifying* role. Insofar as a substance possesses a multitude of powers or properties, there is an open question about how and why these powers or properties are always found together. Why, for example, when an apple is in motion do its redness and sweetness and solidity all move with it? One compelling answer has it that each of these properties is attached to, or inheres in, a given substance. When the substance moves it brings the properties along with it.

[83] It is common to mention Locke in this connection. And Locke's interrogation of the idea of substance is particularly thorough and well-known. But Locke could hardly have been a source for Newton's thinking in *DG*. My suspicion is that Locke developed his views in light of the same currents of thought Newton was responding to in *DG*. Writers like Thomas Hobbes, Henry More, and Samuel Parker had all problematized the idea of substance or our cognitive access to it in one way or another. See, respectively, *CSM*, II, 130; More 1650, page 6; More 1653, page 11; More 1659, pages 10–11; Parker 1666, pages 63–64. See also Blundeville 1599, page 17.

[84] For a helpful enumeration of the tasks undertaken by theories of substance in the context of early modern philosophy, see McCann 2001, pages 87–90.

Given all this, we might wonder about the extent to which Newton's rejection of the idea that bodies have "substances" puts pressure on the *unified* nature of these Newtonian bodies. If Newtonian bodies are just collections or bundles or powers, what is it that keeps them bundled together? More precisely, when impenetrability translates from one part of space to another, why does the power to interact with minds translate along the same path? Call this the Glue Problem for Newtonian secondary substances. What is the *glue* that keeps all the powers attached to one another?[85]

In considering this problem, Kochiras and Downing have arrived at a shared solution.[86] Namely, they invoke the continual action of God to do the needed work. Here is Kochiras:

> These mobile sets of powers must somehow be unified, so as to maintain their characteristic configurations, as they either rest or move through space, and I propose that he [Newton] assigns the task of unifying them to God. The powers are unified and maintained as enduring configurations by God—by ye divine arm, to borrow a phrase that Newton uses elsewhere.[87]

Kochiras is at pains to mention that her solution "falls considerably short of occasionalism," even occasionalism of a "soft" sort.[88] While questions of terminology can always be debated, it is certainly true that the solution does invoke continuous and immediate divine activity that directly produces a huge number of observed phenomena at every moment throughout the entire universe. Using the rubric above, the position falls into category [F]: quasi-occasionalism.

Here again I believe that fully understanding mobility as a genuine power of Newtonian bodies will solve a problem and will offer a solution that precludes the need for high levels of divine involvement in the created world. Note, for example, the way in which Kochiras describes the powers constituting bodies as "mobile" in the block quotation above. This characterization is, in one sense, deeply

[85] In connection with the glue problem, it is intriguing to note that Charleton says three accounts of bodies have been proposed and that the Epicurean one, the one he favors, holds that "by Bodie is to be understood a congeries of figure, magnitude, resistence (or solidity and impenetrability mutual) and gravity." Charleton 1654, page 24. A congeries as defined by the Oxford English Dictionary is "A collection of things merely massed or heaped together."

[86] Kochiras 2013 and Downing 2014, pages 112–114. Howard Stein also considers the Glue Problem and, while his position is slightly harder to interpret, he may also be a proponent of this solution. See Stein 2002, pages 278–279.

[87] Kochiras 2013, page 333. One note on Newton's use of the phrase "ye divine arm": As Kochiras notes, the phrase comes from Newton's correspondence with Bentley, rather than from *DG*. And while Kochiras does not use the phrase to do any substantial load-bearing work in her argument, it is worth emphasizing that Newton invokes the divine arm while describing a type of divine causal activity very different than the type Kochiras advocates for in this passage. Specifically, it refers to a specific one-time action performed by God while creating the solar system, not to a form of universal and continual action in the universe. See *NCS* #399.

[88] Kochiras 2013, pages 335–336.

misleading. One of the powers *is* the power of mobility. The other two powers, considered in and of themselves, are not mobile. They stand in need of some further explanatory posit to make sense of how they might move. This, to my mind, is why Newton introduces mobility as a separate power. It does essential work in the theory. Indeed, my view is that the power of mobility is the glue that keeps the powers of bodies united together.

Understood in this way, mobility is not just a bare power to move. Instead, mobility can be understood as the power to move *some thing* (or some other power). Mobility, so construed, is a sort of "grab-and-pull" power. If this is correct and mobility is the power that ensures impenetrability and mind-interactivity can move through space, then there is no special difficulty in allowing that it is the thing that ensures they move through space together.

The text of *De gravitatione* is less determinate on this issue than one might like. Indeed, it can give the sense that Newton was not even aware of the difficulty. Perhaps this is correct, but if so that would be disappointing. The internal logic of the account in *De gravitatione* demands some kind of answer here. Bodies display a certain kind of unity. This means that Newton will not have given an account of how certain divinely created powers can have all the features and properties of bodies as we know them unless he can say something about how the different created powers are unified. Given this, it is worth attending to the passages in the text that bear on the issue. The most careful definition of bodies just allows that they consist of determined quantities of extension endowed with certain conditions and that one of these conditions is mobility. This is compatible with the interpretation I am supporting, but is hardly a decisive text in favor of it. Slightly more helpful is the initial presentation of mobility in the account:

> If we should suppose that that impenetrability is not always maintained in the same part of space but can be transferred here and there according to certain laws, yet so that the quantity and shape of the impenetrable space are not changed, there will be no property of body which it does not possess.[89]

Here it seems that whatever is responsible for mobility is also responsible for preserving quantity and shape. This indicates that mobility might play something like the unifying role indicated above.

Again, my view is that the issue is not fully determined by the text. This means it is difficult to find passages which clearly support my reading of mobility. But it equally means that it is very difficult to find passages which support the interpretation of Downing and Kochiras. Contextually, however, I think there are two powerful considerations that push back against the idea that Newton understood

[89] *DG* 42.

God to be immediately and continuously responsible for the unified nature of bodies. First, Newton draws a parallel between the powers constituting a body and a substantial form. And a chief task of the substantial form was to account for the kind of organic unity had by a substance. This suggests that the powers themselves have—in some way or another—whatever is needed for unity. Second, the basic framework of bodies in *De gravitatione* involves God creating powers that are designed to recreate all of the observed features of bodies. One of the observed features of bodies is unity. And surely Newton would have allowed that something like the "grab-and-pull" power of mobility that I am advocating for was within the power of God to create. So, at a minimum, there is a version of the account that relieves God of the need to continually unify all bodies through direct action. My hope is that the considerations already adduced in this chapter, combined with those presented in the next section, provide a strong case that this was the version Newton himself endorsed.

5.9. God and the Soul of the World

Further evidence that Newton rejected approaches to nature that were occasionalist or quasi-occasionalist comes from his repeated insistence that God is not an *anima mundi*. While the concept of the world soul has a complex history with roots in ancient Greek philosophy, the basic idea held that there existed a single spirit or force that permeated the universe and animated nature, directing its causal activity.[90] Here is Thomas Stanley's initial presentation of the idea, from his discussion of Thales in *The History of Philosophy*:

> That *the world is animated*, and *that God is the soul thereof, diffus'd through every part, whose divine moving vertue penetrates through the element of water.* Thus explain'd by the *Hermetick* Philosophers; the divine spirit who produc'd this world out of the first water, being infus'd as it were, by a continuall inspiration into the works of nature, and diffus'd largely through, by a certain secret, and continuall act, moving the whole, and every particular according to its kind, is the soul of the world.[91]

Notable here is not just the omnipresence of the world soul and its diffusion through every part of nature, but also its causal contribution to nature. The world soul acts continually to generate motion and the particular behaviors of specific

[90] For helpful background on the concept of a world soul in early modern philosophy, see Peterman 2021. For a helpful discussion from the early modern period, which engages many classical authors, see Burthogge 1699.

[91] Stanley 1655, page 14. Emphasis in original. It is worth noting that Newton owned the 1701 reprint of this book. *NL* #1552.

beings. I suspect Stanley's suggestion that Thales himself identified the world soul with God is on shaky interpretative ground. But identifications of God and the world soul were common enough among others and the view has a certain appeal.

There is a passage from Newton discussing the views of ancient authors that nicely parallels the presentation offered in Stanley.

> Up to this point I have explained the properties of gravity. I have not made the slightest consideration about its cause. However, I would like to relate what the ancients thought about this. Quite apparently the heavens are nearly free of bodies, but nevertheless filled everywhere with a certain infinite *spiritus*, which they call God. The bodies, however, move around freely in this *spiritus*, as a consequence of its forces and natural efficiency they are constantly thrust toward each other, more or less <strongly> in accordance with the harmonic ratios of the distances, and gravity consists in this impact. Some differentiated this spiritus from the highest God and called it the world soul.[92]

A few remarks are needed. The passage describes a certain picture of the universe and its workings. Newton is very careful to specify that this is not his picture; in this passage he claims he has not given "the slightest consideration" to its truth value.[93] Instead, he is merely describing the view of some unspecified ancients. As in Stanley, the view makes a certain entity causally responsible for the continuous production of natural phenomena. And, as in Stanley, it accomplishes this task by being diffused throughout the universe. The final sentence clarifies that some take this entity to be God whereas others take it to be a subordinate world soul. From this we can infer that those who take the entity to be God take God to in some sense be, or to fulfill the functional role of, the world soul.

We are now in a position to see why the issue of the world soul is relevant for the question of Newton's views on divine action in the universe. Simply put, the

[92] Gregory Ms. 247, 14v. Underlining added. Translation taken from Schüller 2001, page 241. See also Newton's discussion of ancient views on the world soul in *Out of Cudworth*. UCLA Clark Library, Ms. Annex 3, Misc. Mss and Broadsheets Box 4, Folder 10, 2r–2v. Levitin 2022, pages 593–598, offers an illuminating and historically sensitive discussion of this passage. I am persuaded by many of the points he makes but hasten to disagree with one of his conclusions, namely that Newton believed "it was imperative to maintain that God was *not* immanent in the world." Page 598, emphasis in original.

[93] In this sense, I disagree with McGuire and Rattansi 1966, which argues that Newton took himself to be following the precise views of the ancients in adopting occasionalist or quasi-occasionalist views. See the criticisms of McGuire and Rattansi in Casini 1984. Further, for all that Newton believed certain of the ancients had gotten right (atomism, heliocentrism, etc.), he certainly believed they had either misunderstood or misrepresented some of the details of gravitational attraction. Note that in the block quotation above—and elsewhere in the Classical Scholia—Newton presents the ancients as claiming that the force of gravitation varies with distance in accord with harmonic ratios, rather than in accord with the inverse square law. For details and discussion of this point, see Guicciardini 2013, pages 61–66. Also notable is the fact that Newton elsewhere indicates that the ancients were either agnostic about or did not agree on the cause of gravitation. See Newton 1728, pages 2–3 (English translation at Newton 1731, page 4). See also CUL Ms. Add 3970, 619r where the "ancient philosophers" affirm that atoms have gravity without "telling us the means" by which it operates, offering only "figures" or analogies.

role played by God in occasionalism seems to be precisely the role of a world soul. Identifying God with the world soul is a move in the direction of occasionalism or at least in the direction of a robust concurrentism. Rejecting such an identification—as Newton does—renders interpretations with very high levels of continuous divine involvement untenable.

Newton was comfortable enough with the idea of a soul being that which gives life and motive force to a body. In a draft letter to Conti, Newton complains that Leibniz "colludes in the signification of words" in part because he calls "those things the souls of men wch do not animate their bodies."[94] But Newton was adamant that God was not the soul of the world.[95]

A particularly strident denial comes in Newton's *Account of the Commercium Epistolicum.* Here Newton is working to emphasize the differences between his own views and those of Leibniz: "The one [Newton] teaches that God (the God in whom we live and move and have our being) is omnipresent, but not as a soul of the world."[96] Newton's contrast here is between his view of God as immanent in the universe and Leibniz's transcendent God, an *intelligentia supramundana.* In this text, Newton allows that Leibniz's God is also not a soul of the world. But his view seems to be that part of Leibniz's motivation for making God transcendent was that there would be no way for God to be in the world without God being a soul of the world. This comes out clearly in the above-mentioned letter to Conti, where Newton complains about the ways in which Leibniz "falls foul upon my Philosophy."[97] Leibniz, according to Newton, "tells you also that God cannot be in the world without animating the world."[98] Or, more plainly in a draft for the *Account of the Commercium Epistolicum*: "He said that God is Intelligentia supramundana because if he were in the world he would be the soul of the world, that is, he would animate the world."[99] The clear suggestion here is that Leibniz is incorrect on this point and that this represents a lack of either imagination or insight on his part. Newton believed that God is in the world but does not animate it.

The line Newton takes against Leibniz here is particularly striking given the latter's charge in the Leibniz-Clarke correspondence that Newton's position was akin to deism. If Newton believed that God was intimately responsible for all actions in nature, if he believed God were anything like a world soul, he might have

[94] CUL Ms. Add 3968, 564r with insertion from 565r. A draft of a letter to Des Maizeaux has subtly different phrasing which, I believe, better captures Newton's actual view. Newton complains that, according to Leibniz: "man himself is a mere machine His body is not actuated by any mind." CUL Ms. Add 3968, 436r. See also 436v.

[95] I thus disagree with the reading offered at Snow 1926, page 209.

[96] *NPW* 167.

[97] CUL Ms. Add 3968, 571r.

[98] CUL Ms. Add 3968, 571r. For more on the internal Leibnizian logic of this complaint see Kulstad 2008.

[99] CUL Ms. Add 3968, 588r. Newton offers a similar formulation at 587v: "He presents that God must be Intelligentia supramundana least he should be the soul of the world." Another similar formulation is at 414v.

seized on this opportunity to set the record straight. Instead, he explicitly rejected the position. Unfortunately, we do not, in this context, get a clear sense of *why* Newton rejects the idea of God as world soul.

Insight into Newton's reasoning can be had from two other texts. The first appears in the General Scholium. The point there is made in the following way:

> He [God] rules all things, not as the world soul but as the lord of all. And because of his dominion he is called Lord God *Pantokrator*. For 'god' is a relative word and has reference to servants, and godhood is the lordship of God, not over his own body as is supposed by those for whom God is the world soul, but over servants.[100]

This passage displays one of Newton's central preoccupations, namely, the danger of interpreting claims about God's moral authority in metaphysical terms. What is centrally important for Newton is God's dominion over all things, and especially over humanity, as ruler and law-giver. Those who make God a world soul conflate this dominion, God's proper dominion, with a metaphysical conception of God as a central animating force.[101]

The second key passage comes from the Queries. It occurs shortly after one of the passages in which Newton identifies space with God's sensorium. The passage rehearses the familiar Newtonian theme, a theme at issue in the *Account of the Commercium Epistolicum* passage considered above, that God's substantial omnipresence in space is what makes God's causal agency in the world intelligible and that God is able to act in the world in a manner similar to the way in which humans act on our own bodies. Nevertheless, we should not be misled:

> And yet we are not to consider the World as the Body of God, or the several Parts thereof, as the Parts of God. . . . they are his Creatures subordinate to him, and subservient to his Will; and he is no more the Soul of them, than the Soul of Man is the Soul of the Species of Things carried through the Organs of Sense into the place of its Sensation.[102]

Although Newton does not use the term "world soul" here, the context makes clear that this constitutes another instance of Newton rejecting the idea that God is the world soul. He does emphasize the subordinate and subservient nature of creatures, but this can be seen as referring to God's having moral authority over

[100] *Principia* 940.

[101] For more on this aspect of Newton's theology, and the general theology undergirding the General Scholium, see Snobelen 2001. See also Snobelen 2005. Hutton 1999 makes the interesting argument that even Newton's Christology reflects this central focus on God's dominion and rulership.

[102] *Opticks* 379. Newton deploys the same analogy in a draft of a letter to Des Maizeaux and in a draft of the *Account of the Commercium Epistolicum*. See, respectively, CUL Ms. Add 3968, 436r and 587v.

them. Recall from above Newton's comfort with the way in which the human soul can be understood as that which animates the body. His suggestion here is that while the soul of humans animates the body, it clearly does not play this role for the visible species received in perception. They are, in some sense, external objects brought into the body. In the same way, created substances are not part of a thing fundamentally animated by God. And while God's intimate presence guarantees that God has knowledge of them, it is does not indicate that God constantly acts on them.

Thus far, we have seen Newton charting a fairly clear course on the issue of God as a world soul. Newton wants to emphasize God's intimacy with the world. But, as claimed above, this intimacy is achieved through God's immediate presence to and moral dominion over all things. It is not achieved through God's continuous causal involvement in nature. And this is what lies behind Newton's claim that God is not the soul of the world. Of course, God's activity was needed in creation. And God might, in keeping with certain providential interests, causally intervene in the universe on certain occasions. But these instances mark a departure from the ordinary. Insofar as this is correct, the occasionalist and concurrentist readings of Newton seem to face a major objection. Newton's repeated insistence that God is not the soul of the world is very difficult to square with any view that has him continually playing a role in all the causal processes of nature. As Newton wrote to Burnet: "Where natural causes are at hand God uses them as instruments in his works."[103]

There is one Newtonian text which might be seen as running counter to the interpretation developed thus far in this section. The passage occurs in *De gravitatione*, making it particularly important to consider. Further, Lisa Downing has appealed to the passage to argue for an interpretation which emphasizes God's continuous activity in the world.[104] Here is the relevant text:

> Rather, if anyone should think it possible that God may produce some intellectual creature so perfect that it could, by divine accord, in turn produce creatures of a lower order, this I submit does not detract from the divine power, it posits an infinitely greater power, by which creatures would be brought forth not only directly but by other intermediate creatures. And so some may perhaps prefer to posit a soul of the world created by God, upon which he imposes the law that definite spaces are endowed with corporeal properties, rather than to believe that this function is discharged by God. To be sure, the world should not be called the

[103] *NCS* #247. It should be noted that this line to Burnet is far more damaging to occasionalist interpretations than to concurrentist ones. In connection with this comment to Burnet, it is also worth noting that Newton even developed an account of miracles which did not require direct divine action. A particularly clear and concise statement of this understanding of miracles can be found in Lehigh Ms. ALS Newton I. 17xx. For discussion, see Harrison 1995.

[104] Downing 2014, pages 112–114. A similar reading appears at McGuire 1968a, page 156.

> creature of that soul but of God alone, who creates it by constituting the soul of such a nature that the world necessarily emanates [from it]. But I do not see why God himself does not directly inform space with bodies.[105]

Downing reads the passage as claiming that "the posited world soul would be continually carrying out some function that, instead, God might directly (and continually) perform."[106] Downing is clearly right that Newton thinks God could, if God wished, delegate certain actions to a world soul but that supposing God does so is extraneous. God would act more parsimoniously by simply doing things Godself.[107] My disagreement with Downing has to do with whether anything in this passage pertains to continuous action, rather than the initial act of creation. Downing is certainly correct that *De gravitatione* aims to give an account of the nature of bodies, and not just the way in which they are created. But the specific passage in question seems squarely focused on the creation of the world and not on its subsequent maintenance. I see no reason why it should be taken as offering any indication about what is required for the operations of bodies after they are created. At a minimum, the passage could at least be read as neutral on the question of God's continuous involvement.

Early modern thinkers understood the world soul as something exerting causal agency in order to produce ordered and harmonious activity in the natural world. Newton consistently and stridently denied that God should be understood as, or as being like, a world soul. Even in contexts where he was being accused of deism and in which it would have therefore been useful for him to at least cede some ground to the idea that God was a world soul, Newton was steadfast in rejecting the view. This section has argued that this presents a major difficulty for interpretations that adopt some form of the positions characterized as [E], [F], [G], and [H] above. At a minimum, it is incumbent on proponents of those views to offer some alternative interpretation of what Newton might have been meaning to do in denying that God was a world soul and to show how a denial of God as a world soul could be compatible with concurrentism or some version of occasionalism.

5.10. Conclusion

This chapter began with the observation that Newtonian bodies, as secondary substances, are defined by their dependence on God. We are now in a position to clarify what that dependence consists in. They depend on God in the sense that

105 *DG* 45.

106 Downing 2014, page 114.

107 Given this strident stance in *DG*, it is interesting to see Newton claiming, in a later text from the 1710s, that "God does nothing by himself which he can do by another." Yahuda Ms. 15.4, 67r. See also the canceled portion of the text at Keynes Ms. 9, 4r.

God must create them in order for them to exist. And they must be conserved by God at every moment. But this is the extent of their dependence. Bodies have (or, better, are) genuine causal powers; they are substances and Newtonian substances act. This means that Newtonian bodies enjoy a far higher degree of autonomy than is commonly recognized.

We also now know what Newtonian bodies are. They are collections of spatially extended powers with determinate quantities. Specifically, they are the powers of mobility, impenetrability, and mind-interactivity. Knowing this lets us understand another sense in which bodies depend on God. They depend on God not just for the fact *that* they are (and continue to be) but also for *what* they are. On a traditional Scholastic picture, God creates a being by actualizing some essence. The essence then determines the other features or powers of that being. This is not Newton's picture; bodies have no central essence. They just are collections of powers. Because bodies fundamentally are collections of powers, and because God freely choses which powers to bundle together in creating bodies, bodies depend on God with respect to what they fundamentally are.

6
Later Developments in Newton's Theory of Bodies

6.1. Introduction

The previous chapter reconstructed the theory of bodies developed in *De gravitatione*. As argued in Chapter 4, that theory forms the basic core of Newton's approach to bodies. But, as also argued in Chapter 4, while Newton's commitment to the basic idea of *De gravitatione* is consistent throughout his career, the precise details do undergo some development. The aim of this chapter is to explore that development.

My approach to these issues differs from that of some other important commentators. Katherine Brading has developed an influential interpretation according to which *laws* rather than *powers* are at the center of Newton's conception of body. Zvi Biener and Eric Schliesser have also given a certain kind of priority to laws in Newton's thinking about fundamental metaphysics. And Zvi Biener and Christopher Smeenk see a critical shift between the account of bodies in *De gravitatione* and the *Principia* relating to the introduction of mass. My view is that it is possible to tell a story of Newton's thinking on these issues throughout his career that foregrounds a kind of continuity. There is no radical break with the basic theory in *De gravitatione*, understood in the way I have interpreted it. I will first sketch out my positive interpretation of the way Newton's theory develops before offering a critical analysis of readings that contrast with my own.

6.2. The Positive View in Connection with *De gravitatione*

Here is the basic idea behind my positive proposal: Newton was attracted to the theory of bodies in *De gravitatione* in part because it provided a conceptual and metaphysical framework in which empirical investigations could take place. This thought should, in some sense, be unsurprising. After all, in the course of concluding his negative arguments against the Cartesian account of motion and transitioning to his positive account of "extension and body" and "how they differ from each other," Newton tells us that the account will be useful precisely because it will "lay truer foundations of the mechanical sciences."[1] My understanding is that

[1] *DG* 35. Palter 1987, page 389 helpfully flags that this point "cannot be too strongly emphasized."

Newton's Metaphysics of Substance. Patrick J. Connolly, Oxford University Press. © Patrick J. Connolly 2025.
DOI: 10.1093/9780191959240.003.0006

Newton was committed to the idea that bodies were collections of certain powers such as mobility, impenetrability, and mind-body interactivity, but he felt that precise characterizations of those powers required careful observation, experiment, and analysis. That was the work of natural philosophy.[2]

Before proceeding further, it is necessary to say something about powers, forces, and the relationship between the two. Quite obviously, Newton's physics is framed in terms of *forces*. It tells us how bodies will behave when subject to various forces. For example, from the Second Law we learn that "A change in motion is proportional to the motive force impressed and takes place along the straight line in which that force is impressed."[3] But the *Principia* does precious little to tell us about the metaphysical nature of these forces. Accordingly, there has been considerable debate about whether or not the forces characterized in the *Principia* should even be thought of as having a metaphysical nature. Perhaps they are best understood in strictly mathematical terms or are only meant to be used instrumentally as tools for the completion of various calculations. Following Janiak, I think approaching forces in a metaphysically neutral or positivist way would be a mistake.[4] The forces of Newtonian physics are meant to be "real" in the sense that they pick out features of the genuine causal structure of nature. As characterized in Definition 5 of *De gravitatione*, "force is the causal principle of motion and rest."[5]

But forces are not free-floating constituents of the Newtonian universe. Forces need to have grounds; they have to come from somewhere. We need something that explains why a body will exert a force in a certain way or to a certain degree in some conditions but in a quite different way or to quite a different degree in other conditions. The conceptual resource we have for this is a power. Reflection on the fact that powers are dispositions can help to make this clear. Dispositions can either be dormant and unactivated or "triggered" such that they manifest. The forces in Newtonian physics, I claim, are the activations or manifestations of underlying powers.[6] Understood in this way, we can fix a clear relationship between the powers that I claim are the metaphysical constituents of Newtonian bodies and

[2] This position has affinities with the ones developed at Stein 2002, pages 288–289 and Janiak 2008, page 118. For helpful background to this view in Janiak's thought, see pages 13 and 44–45. As noted in the introduction, I am skeptical of seeing too much unity across all aspects of Newton's work. But it may be that the general picture outlined here does have parallels in other areas of Newton's thought. Consider the following claims from Buchwald and Feingold 2013, pages 196–197: "Once Newton had formulated the essential structure of his theory of the evolution of civilization, he treated it in much the same way that he had comparable issues in mechanics and optics. In those situations, as we have seen, Newton developed a basic structure, one that he held to firmly. That structure was in turn compatible with a spectrum of possible values for the variables that appeared in it, or even for the presence of specific terms.... But the values of the distinct terms had to be obtained both from theory and from experiment."

[3] *Principia* 416.

[4] See Janiak 2007; Janiak 2008, ch. 3; Janiak 2013.

[5] *DG* 50.

[6] For insightful analyses of a distinction along these lines and its place in Newton's thought, see McGuire 2017 and Janiak 2023.

the forces that are the subject of Newtonian physics.[7] Powers constitute bodies. The exercise of those powers results in force.

With this understood, we can return to the basic idea linking up the theory of bodies in *De gravitatione* with Newton's natural philosophy. The theory of bodies has it that bodies are composed of various powers and the task of natural philosophy is to empirically investigate these powers with the aim of offering fuller and more detailed characterizations of them. To begin getting a sense for this, we can consider the way in which saying that bodies have a power of impenetrability underdetermines what bodies are like and how they behave. A simple thought experiment can help to make this clear. Imagine two bodies of equal size moving toward one another at equal speeds. What will happen when they meet? If impenetrability is characterized as nothing other than the ability to prevent co-location with other bodies, then we have no way giving a full and complete answer to this question.[8] "Mere" impenetrability is compatible with them coming to rest adjacent to one another, with them repelling one another and each reversing direction, or indeed with any other state of motion or rest the bodies might exhibit provided they occupy distinct, non-overlapping spaces at each individual moment. But the fact is, in our world, bodies behave in very specific ways during collisions. So, an understanding of impenetrability as the bare power to prevent co-location is inadequate. More precisely, it is underspecified in the sense that "the power of impenetrability" actually picks out a set of distinct powers, some of which would produce some types of activity in bodies and others of which would produce quite different ones. Thus, in his most careful treatment of impenetrability, Newton notes that this power of bodies not only prevents bodies from coinciding, but it prevents them from coinciding such that "oppositions obstruct their mutual motions and they are reflected in accord with certain laws."[9]

Seen in this way, a more detailed analysis of the powers that constitute bodies is needed. If the explanatory ambition of *De gravitatione* to "describe a certain kind of being similar in every way to bodies" is to be fulfilled, then the powers must be described in a way that accounts for all of the phenomena bodies display.[10] On my interpretation, Newton thought the best and most precise characterization of these powers could be offered using mathematics.[11]

For the case just discussed, the power of impenetrability, Newton's understanding was that precise mathematical characterizations of the power had been offered prior to the *Principia*. Faced with the manifest empirical inadequacy of

[7] Of course, one difficulty in parsing these issues stems from the fact that *vis* in Latin has a semantic range broad enough to encompass both things like applied forces and things like intrinsic powers or capacities. While it is almost always translated as "force," it is often worth asking whether and where "power" could be an acceptable substitution.

[8] Chen 2020, page 194, offers a helpful formulation of this problem. See also Janiak 2008, page 107.

[9] *DG* 13.

[10] NPW 42.

[11] In this sense, I agree with the position articulated at Janiak 2008, page 57.

Descartes' laws describing collisions, the Royal Society wanted, in the words of Oldenburg, that "those in the Society who had meditated on the natural character of the motions to be investigated be asked to produce their meditations and discoveries on the subject."[12] As is well-known, this produced responses from Christopher Wren, John Wallis, and Christiaan Huygens.[13] On Newton's view, these three individuals, who were "easily the foremost geometers of the previous generation," had "found the rules for the collisions and reflections of hard bodies."[14] On the view being developed here, they can be understood as offering mathematical characterizations of the power of impenetrability.[15]

If one held a broadly mechanistic view of nature, then describing what happened in collisions would be sufficient to account for the naturally produced motions of bodies. For example, Thomas Sprat, in writing about Wren's account of collisions says that he had proposed:

> [T]he Principles of all *Demonstrations* in *Natural Philosophy*: Nor can it seem strange, that these *Elements* should be of such Universal use; if we consider that *Generation*, *Corruption*, *Alteration*, and all the Vicissitudes of *Nature*, are nothing else but the effects arising from the meeting of little bodies.[16]

Newton, certainly by the time of the *Principia* if not well before, rejected this restrictive understanding of change in nature. He was interested in more than motion arising from collision or from impact. Put in the language of his theory of bodies, he was unwilling to allow that a full characterization of the power of impenetrability would be sufficient to provide a full characterization of the power of mobility. The two powers could not be collapsed into one another. Put in a more physical register, Newton believed that motions were generated by forces impressed on bodies. Some of these forces were impressed as a result of collisions between bodies (as a result of impenetrability). But there were other forces as well.

[12] Murray, Harper, and Wilson 2011, page 154.

[13] For the text of their contributions and helpful discussion, see Murray, Harper, and Wilson 2011. For an illuminating discussion of this episode in the context of the Royal Society, see Jalobeanu 2011. Relatedly, see Kemeny 2017.

[14] *Principia* 424. There is a slight puzzle surrounding the presence of "hard" in this sentence. The first two editions of the *Principia* printed "duorum corporum,", indicating that Wren, Wallis, and Huygens had dealt with collisions between two bodies. In the third, this appeared as "durorum corporum," picking out collisions between hard bodies. Given that Wren and Huygens offer rules for perfectly elastic collisions and only Wallis gives rules for perfectly inelastic collisions, the description of all three as dealing with collisions of hard bodies is somewhat surprising. Thus, "durorum" seems to me to have been likely a typesetting error rather than a change introduced by Newton himself. At a minimum, the phrase goes without notice or instruction for change in Newton's annotated copy of the second edition, which was used by Pemberton in preparing the third. See CUL shelfmark Adv.b.39.2. In any event, later in the Scholium, Newton makes it clear that he had given thought to both elastic and inelastic collisions.

[15] As it happens, issues surrounding collisions were not settled as decisively as Newton indicates here. For an excellent treatment of problems surrounding collision in the post-Newtonian period, see Brading and Stan 2023. See also Brading and Stan 2021.

[16] Sprat 1667, page 312. Quoted by Jalobeanu 2011, page 105.

This meant, *pace* Sprat, that a full characterization of the power of mobility required more than Wren's characterization of impenetrability and its effects. Thus, Newton was motivated to search for a precise mathematical characterization of the power of mobility. On my account, the *Principia*'s famous three laws of motion provide such a characterization.

6.3. Further Development of the Theory

Up to this point, the views being attributed to Newton have consisted of a refinement or further specification of the powers in his theory of bodies as outlined in *De gravitatione*.[17] But, as the years progressed, Newton became increasingly convinced that there are important physical phenomena displayed by bodies which are not obviously explicable by appeal to even well-defined powers of mobility and impenetrability.[18] The cohesion of fundamental particles that produces macro-sized objects is one.[19] Phenomena arising from magnetism, electricity, and chymical processes are others.[20] It is, of course, widely recognized that Newton came to introduce non-mechanical *forces* as a result of these insights. The correlate to this, understood within the context of his fundamental metaphysics, is that Newton came to believe additional distinct *powers* of bodies could be posited to explain these forces and, ultimately, the relevant phenomena.[21] His aspiration, then, was to offer a full mathematical characterization of those powers. In one notable case, that of gravitation, he succeeded. In many others, he was ultimately unsuccessful. But examining even these failures and what Newton says about them will be instructive. It will help to illuminate what he saw as the task of natural philosophy vis-à-vis his commitments regarding the metaphysics of bodies.

The Preface to the first edition of the *Principia* provides a good starting point for the discovery of this research program in Newton's texts. Here, famously, we learn that "the basic problem [*lit.* whole difficulty] of philosophy is to discover the

[17] Strictly speaking this is incorrect, as I have said nothing about mind-interactivity. Even here, however, we might see some evidence that Newton desired to provide a precise mathematical characterization of the power. See *CPQ* 431–443 and *NCS* #588.

[18] Henry 2023 argues plausibly that *De aere et aethere* marks an important transitional point in this process. See also Benítez 2006.

[19] Ultimately this cohesion, and variations in its force between the particular fundamental bodies composing individual macro-bodies, might also be responsible for the phenomena associated with elastic collision. At a minimum, Newton clearly attributed the "softnesse" or "springynesse" of some macro-bodies to the fact that they were composed of smaller bodies. See "Some Observations about Motion" in *NCS* #349.

[20] In connection with this, it is worth noting that Isaac Barrow had criticized Cartesian mechanism as inadequate to account for phenomena similar to those that preoccupied Newton. See Kargon 1966, pages 78–79 and Stewart 2001.

[21] In this sense, Newton's project can be seen as broadly continuous with some mainstream trends in early modern English corpuscularian theory. See Henry 1986; Guicciardini 2018, page 109; Belkind 2020, sec. 4.2.

forces [*vires*] of nature from the phenomena of motions and then to demonstrate the other phenomena from these forces."[22] Newton brags about having successfully done this for gravitational forces and expresses the desire to do something similar for other phenomena in nature before saying the following:

> [M]any things lead me to have a suspicion that all phenomena may depend on certain forces by which the particles of bodies, by causes not yet known, either are impelled toward one another and cohere in regular figures, or are repelled from one another and recede.[23]

Newton then claims that previous efforts at natural philosophy have failed because the forces in question have not yet been found and characterized before registering his hope that the *Principia* will improve the situation. Newton is arguing here that there must be additional attractive powers comprising bodies that account for phenomena like magnetism and cohesion and additional repulsive powers comprising bodies that account for various chymical phenomena. In this sense, to borrow a phrase from John Henry, the *Principia* was "only a preliminary sortie."[24] The task of natural philosophy going forward would be to uncover these other powers and to discern how they work.[25]

We can focus first on the successful case: gravitation. Robert Hooke offers a helpful route into the consideration of this Newtonian program. In his 1674 *Attempt to Prove the Motion of the Earth by Observation,* a book owned by Newton, he allowed that a successful system of the world could be constructed on the basis of three principles.[26] The first was universal attraction: all bodies attract all other bodies (at least up to a certain distance). The second was the law of inertia. Hooke's description of the third is particularly interesting:

> The third supposition is, That these attractive powers are so much the more powerful in operating, by how much the nearer the body wrought upon is to their own Centers. Now what these several degrees are I have not yet experimentally verified; but it is a notion, which if fully prosecuted as it ought to be, will mightily assist the Astronomer to reduce all the Cœlestial Motions to a certain rule, which I doubt will never be done true without it.[27]

[22] *Principia* 382.

[23] *Principia* 382. A much clearer and more detailed version of these claims can be found under "Hypoth 2" in CUL Ms. Add 3970, 338r–338v. See also the expansive treatment of this general theme in the draft preface and draft conclusion to the first edition of the *Principia* found, respectively, at CUL Ms. Add 3965, 620r–620v and at CUL Ms. Add 4005, 25r–27r (see also 30r–37r), both of which are transcribed and translated at Hall and Hall 1962, pages 302–308 and pages 320–347.

[24] Henry 1988, page 127.

[25] Note the proposal here runs counter to the "law-first" epistemology and methodology attributed to Newton in Hazelwood 2023.

[26] *NL* #794.

[27] Hooke 1674, page 28.

Here Hooke allows that there is some power of attraction explaining the motions of bodies. What remains to be done is to investigate the power more fully in order to offer a precise description of it. The power gets stronger as bodies get closer. But Hooke is unsure about to what degree. This is what remains to be experimentally verified. I see Newton as adopting a similar approach. The first stage is to recognize that one of the powers in nature is that of gravitational attraction.[28] The next task is to use careful observation and experiment to supply a precise mathematicized characterization of this power. Newton can then be understood as succeeding in this second step in a way that Hooke (at least in 1674) did not. He supplied a proof for the "certain rule" that Hooke sought in the block quotation above: the inverse-square law.[29]

Newton's efforts to differentiate and fully characterize other powers, specifically other powers of attraction and repulsion, are on display in a number of texts. Some, like the Queries, are familiar. Others, like draft conclusions for the first and second editions of the *Principia*, are perhaps less well-known.[30] But the general program is clear enough across a number of texts. A useful articulation of it can be found in draft of a Preface for the second edition of the *Principia*. After noting the improvements he had made to the explanations of phenomena caused by gravitational attraction, Newton says the following:

> [I]t remains to investigate the remaining attractive forces, that is to say the electric force and the force of magnetism, and to discover their laws and their effects upon the motions of the least of bodies[31] [as manifested] in solution, fermentation, vegetation and similar processes.... Thus philosophers may see a broad area lying before them in which they may exercise their wits.[32]

A sketch of what this might look like can be found in Newton's comments about magnetism in Book III, Proposition 6, Corollary 5 of the *Principia*:

[28] Note that the power is a dispositional one, it is the power a body has to attract other bodies should any be present. This should suffice to account for the sense in which gravitational attraction is relational. I am therefore sympathetic to the critique of Schliesser 2011b offered in Henry 2024a. For a somewhat different take on gravitation's status as relational, see Miller 2009. See also Tuchańksa 1988, pages 49–51.

[29] Of course, subsequent to the 1674 *Attempt*, Hooke does find and advocate for something like the inverse-square law. See *NCS* #239 and #240. This leads to his priority dispute with Newton. For details and some helpful, though opinionated, background see Kollerstrom 1999. For more recent perspectives, see the papers making up the 2005 forum "Newton vs. Hooke on Gravitation": Guicciardini 2005; Nauenberg 2005; Gal 2005; Bertoloni Meli 2005.

[30] See Hall and Hall 1962, pages 320–347 and *Principia* 287–292.

[31] The Latin here has "minimorum corporum" but it is worth noting that Newton had first written and crossed out language that was more obviously atomist: "minimarum particularum materiæ corporæ"—the smallest particles of bodily matter.

[32] *NCS* #829a.

> The force of gravity is of a different kind from the magnetic force. For magnetic attraction is not proportional to the [quantity of] matter attracted. Some bodies are attracted [by a magnet] more [than in proportion to their quantity of matter], and others less, while most bodies are not attracted [by a magnet at all]. And the magnetic force in one and the same body can be intended and remitted . . . and this force, in receding from the magnet, decreases not as the square but almost as the cube of the distance, as far as I have been able to tell from certain rough observations.[33]

We see a few things here. First, Newton is committed to the idea that there is a magnetic force in (some) bodies that explains certain phenomena of attraction. Second, Newton is working to establish that this power is different in kind from the one responsible for gravitational phenomena. Put differently, we do not violate the first of the 'Rules for Study of Natural Philosophy' in positing it; we have good reason to think gravitational and magnetic phenomena cannot be explained by a single power. Finally, we see an interest in offering a precise mathematical characterization of the power. There is an effort to determine whether and how the force of the power relates to the quantity of matter of the body that has it. And there is a rough pass at something like an inverse-cube law to characterize the force of the power over distance.

While this basic Newtonian program succeeded in the case of gravitation, it ultimately failed in the case of other powers.[34] Newton was correct in his claim that: "To explain all nature is too difficult a task for any one man or even for any one age."[35] At the end of the General Scholium he writes that he might have added more about the "force and actions" by which particles attract one another at close range and adhere to compose solid composite objects and more about electrical powers by which a variety of phenomena come about. But he holds off. The reason is that he is in the same position as Hooke was in with respect to gravitational attraction. He writes that "there is not a sufficient number of experiments to determine and demonstrate accurately the laws governing the actions of this spirit."[36] A similar attitude is on offer in the Queries. Query 31 begins by asking: "Have not the small Particles of Bodies certain Powers, Virtues, or Forces, by which they act at a distance . . . upon one another for producing a great Part of the Phænomena of Nature?"[37] He offers initial support for the idea through an appeal to the already recognized powers of gravity, magnetism, and electricity and suggests that

[33] *Principia* 810.

[34] Ducheyne 2012, ch. 4, sec. 9 is helpful on this point.

[35] CUL Ms. Add 3970.3, 479r. See also CUL Ms. Add 3970, 244v: "Many experiments are wanting for completing the Analysis of this part of Nature [optics] & coming to a clear & distinct knowledge of all the causes of these things, many more for perfecting the Analysis of all Nature & making a full & clear discovery of all the first Principles of Natural Philosophy. To compass this is a work which requires many heads & hands & a long time."

[36] *Principia* 944. See also his similar characterization of the situation at *NPW* 165 and at CUL Ms. Add 3968, 586r.

[37] *Opticks* 350.

these three "make it not improbable that there may be more attractive Powers than these."[38] While Newton nicely illustrates the ways in which such forces might give rise to various phenomena, it is clear he also stops short of offering a full taxonomy of such forces and mathematical characterization of them. Indeed, Newton's failure to offer such a characterization of the relevant powers might even help to explain the need for the Queries, a section of the *Opticks* with more relaxed epistemic standards.

6.4. An Objection

Lisa Downing, in a discussion of Janiak's argument for the reality of Newtonian force, notes something interesting and prima facie puzzling. Although Downing is sympathetic to Janiak's treatment, she highlights the "difficulty" of the fact that "Newton writes of gravity as both being a cause and as having a cause."[39] On Janiak's view, Downing indicates, only the former really makes sense.

I believe we can make sense of gravity both being and having a cause. We can do so by invoking the distinction between powers and forces discussed above at the beginning of Section 6.2. The solidity of the wall explains why the tennis ball I actually threw at it bounced back. But it also explains why if I *were* to throw a golf ball at it the golf ball *would* bounce back as well. In the former case, we pick out a force the wall exerted. In the latter, we pick out a power it has to exert forces. The force I exert on the door is the cause of its opening, it is sufficient to explain why the door opened. But that force itself also has a cause. My muscular arrangement in combination with other features of my physiology makes it such that I can exert force on doors; I have the power to push on doors. Similarly, Newton can claim that he has "explained the phenomena of the heavens and of our sea by the force of gravity."[40] And he can claim that "gravity really exists and acts according to the laws that we have set forth."[41] He can claim these things with respect to gravitational forces. And he can do so while still purporting to remain agnostic about the *power* that gives rise to those forces. So, he is still entitled to claim that he "has not yet assigned a cause to gravity."[42] This, at least, is one way of making sense of Newton's otherwise curious stance that gravity both is and must have a cause.

Suppose one accepts the position just outlined. Unfortunately, the difficulties—at least for my interpretation—are not at an end. I have said that Newton took gravitational forces to be a key cause of various phenomena (the planets having elliptical orbits, the motion of the tides, apples falling to earth, etc.). I have also said that, for Newton, one of the powers constituting bodies is the power of gravitation,

[38] *Opticks* 351.
[39] Downing 2011, page 125.
[40] *Principia* 943.
[41] *Principia* 943.
[42] *Principia* 943.

which itself is the cause of the gravitational forces. The worry for my interpretation is that this second claim seems to be at odds with a number of (sometimes quite forceful) Newtonian pronouncements in which he insists that he is agnostic regarding the cause of gravitational forces. Here is a sampling:

> I have not yet assigned a cause to gravity.[43]

> . . . the cause of gravity is what I do not pretend to know, and therefore would take more time to consider of it.[44]

> [writing about himself in the third person] . . . he said that for want of a sufficient number of experiments, he forbore to describe the laws of actions of the spirit or agent by which this attraction is performed. And for the same reason he was silent about the cause of gravity.[45]

On my interpretation, the objector presses, this is not at all what Newton ought to have said. He ought to have simply owned the idea that the cause of gravity is one of the constituent powers of bodies in just the same way that the cause of a force impressed in a collision is the power of impenetrability that partially constitutes bodies.[46] But he does not do this; he claims to be entirely ignorant of this cause.

This is a serious objection. I propose two ways of addressing it. The first relates to Newton's intense aversion to public dispute. The second relates to a fundamental epistemic issue facing the Newtonian theory of bodies as I have reconstructed it.

Newton's early papers on light faced spirited opposition, notably from Robert Hooke and a number of Jesuits based on the Continent. This opposition, at least in the case of Hooke, caused Newton to refine his views or at least to offer alternative formulations of them that are quite revealing. It is clear, however, that Newton found the episode distressing. I think it is fair to say that it generated in him a deep desire to avoid public dispute on issues of natural philosophy. For instance, he wrote to Oldenburg during this period asking to resign his membership in the Royal Society and had to be coaxed back.[47] And, employing a regrettable metaphor, he wrote to Halley in 1686 with the complaint that: "Philosophy [meaning natural philosophy] is such an impertinently litigious Lady that a man had as good be engaged in Law suits as have to do with her. I found it so formerly & now I no sooner come near her again but she gives me warning."[48] Newton was simply uninterested in pursuing certain issues in the court of public opinion.

[43] *Principia* 943.
[44] *NCS* #399.
[45] *NPW* 165–166.
[46] On this point, see Connolly 2024, in which I argue that Newton does, in fact, come quite close to saying precisely this.
[47] *NCS* #101 and #102.
[48] *NCS* #288.

Given this, it is unsurprising that the first edition of the *Principia* itself is relatively silent on metaphysical issues and seeks to operationalize key concepts like body and attraction. An early review in the *Journal des Sçavans* picked up on this feature of the book.[49] But even Newton's careful framing was unable to rescue the text from critical interrogation, especially critical interrogation of gravitation as a non-mechanical force. In response, Newton adopted a defensive posture. He argued that he had established the existence of gravitational forces and given solid reasons for denying that gravitation could be understood mechanically. Beyond this, he did little more than point critics toward the *Principia*'s claim that it does not consider the "physical causes and sites of forces."[50] Had Newton publicly owned the position I claim that he held, or one like it, it is hard to imagine that his doing so would have done much to satisfy his opponents. Quite the opposite, it would have only generated more, and more vociferous, disputes. To be clear, none of this is positive evidence that Newton understood gravitation to be one of the powers constituting bodies, as I suggest he did. But it does show that we have some positive reason for thinking that *if he did* understand gravitation to be a fundamental power of bodies, he might have been highly motivated to keep the view under wraps.

Here is a second—somewhat more principled—way of responding to the objection. The basic strategy behind Newton's theory of bodies is to find some number of powers capable of accounting for the features had by bodies. Here are two reasonable constraints on an approach to bodies of this type. First, the powers should be heterogenous. No power should duplicate the work done by another power. Second, no more powers should be posited than are needed to account for the phenomena bodies actually display. These two constraints have obvious resonances with the first two of the Rules for the Study of Natural Philosophy.

The suggestion above was that Newton came to see the broadly mechanistic version of bodies outlined in *De gravitatione* as inadequate. More powers were needed. But the positing of a new type of constitutive power for bodies required a certain amount of epistemic due diligence. Quite obviously, Newton had worked to show that gravitational forces could not be accounted for by the same power that accounted for impenetrability.[51] And we saw above considerations designed to show that gravitation and magnetism have divergent properties. But, Newton may have wondered, might gravitational phenomena be tied up with something else such that positing a gravitational power that did not also account for, say, certain electrical phenomena would be overly hasty? Seen in this light, the trick is to posit just enough powers without positing too many. And one might think that further research was called for before decisively ruling on the issue.

[49] On early reviews of the *Principia*, see Cohen 1990.
[50] *Principia* 407. See, e.g., *NPW* 165–166.
[51] *Principia* 943.

Some of Newton's remarks on occult qualities show that he was sensitive to these sorts of considerations. Here is a passage from a short note on miracles: "by giving this name [i.e., occult qualities] to the specific qualities of things, a stop has been put to all enquiry into the causes of those qualities."[52] The idea expressed here can be productively read in combination with one from the Queries:

> *Aristotelians* gave the Name of occult Qualities . . . to such Qualities only as they supposed to lie hid in Bodies, and to be the unknown Causes of manifest Effects: Such as would be the Causes of Gravity, and of magnetick and electric Attractions, and of Fermentations . . . . Such occult Qualities put a stop to the Improvement of Natural Philosophy . . . . To tell us that every Species of Things is endow'd with an occult specifick Quality by which it acts and produces manifest Effects, is to tell us nothing; But to derive two or three general Principles of Motion by Phænomena, and afterwards to tell us how the Properties and Actions of all corporeal Things follow from those manifest Principles, would be a very great step in Philosophy, though the Causes of those Principles were not yet discover'd.[53]

The Aristotelians erred because they posited too many explanatory principles. Each thing, or each action, was assigned a principle to explain it. On Newton's view, simplifying somewhat, they would posit a *tidal* quality to explain the motions of the tides, an *elliptical* quality to explain the orbits of the planets, and a quality of *heaviness* to explain the motion of falling apples toward the earth. If one were to rest content with these explanations, then not only would one learn nothing, because these look like dormitive virtue explanations, one would also be insufficiently motivated to investigate the relevant phenomena further and to discover that they are all in fact related and can all be explained with reference to a *gravitational* quality.[54]

Of course, in the block quotation above Newton's focus is not on the powers that are constitutive of bodies. But the same form of reasoning transfers quite cleanly to the project of constructing a theory of bodies using heterogenous powers. Positing a unique power every time we encounter some new phenomenon will quickly render the project hopeless. It gives up on the possibility of using unification to generate explanatory power. New powers should only be admitted when we are

[52] Lehigh MS ALS Newton I. 17xx.

[53] *Opticks* 377.

[54] In connection with this, it is interesting to note that the eighteenth-century Newtonian Colin MacLaurin criticized occasionalists along similar lines; concluding that God was immediately responsible for all observed phenomena would, allegedly, remove the motivation for natural philosophical research. MacLaurin 1748, pages 388–389. For some discussion, see Connolly 2016.

confident that they are needed. We ought to do as much as possible with as little as possible.[55]

Did Newton believe that, alongside the powers of mobility, impenetrability, and mind-interactivity, there was a power of gravitation that helped to constitute bodies? I have said, possibly against the weight of considerable textual evidence, that the answer is yes. But the yes is somewhat qualified. I also take seriously Newton's claims that gravitation, magnetism, fermentation, cohesion, and other phenomena should be open to a great deal of further empirical investigation. And, in particular, it seems to me that Newton thought more work was needed to ensure that the same power responsible for gravitation was not also responsible for, say, electricity or short-range attraction.

6.5. Short-Range Attraction and Chymistry

Newton's engagement with chymistry has long been of interest to scholars and has contributed to the debates discussed in the introduction regarding the unity or otherwise of Newton's thought.[56] One might reasonably wonder whether this in-depth and decades-long interest in chymistry is of special import for understanding his views on the character of short-range attractive forces and his attempts to give precise mathematical treatments of them.

There are a number of immediate obstacles to learning about Newton's theoretical approach to short-range forces by attending to his chymical practice. The first is perhaps unsurprising. Many early modern chymists were committed to a certain form of privacy. The secrets of the art were meant only for the adepts and the properly initiated. And chymists took special efforts to make sure public or published statements revealed only partial information and were cloaked in language comprehensible only by other, sufficiently worthy, chymical practitioners.[57] A second obstacle stems from the fact that most—indeed, the bulk—of Newton's chymical activity was intensely *practical* rather than theoretical.[58] And a third difficulty relates to the fact that Newton's views in the chymical arena underwent various shifts and evolutions throughout the course of his career. For example, in an important

[55] An extreme version of this thought can be found in the work of Roger Boscovich. Drawing, in part, on work by Newton, he posited a single power of bodies by which they alternatively attracted and repelled other bodies at greater and lesser distances. The idea was to use this single power to explain a wide range of phenomena ranging from gravitational effects to the cohesion of bodies and much in between. At least this is one prominent reading. For a more nuanced and metaphysically cautious interpretation and a discussion of figures with related views, see Guzzardi 2020, chs. 2–4.

[56] For a dependable guide through the sometimes murky waters of Newton's chymistry, see Newman 2019. For shorter overviews, see Figala 2002, Newman 2002, or Newman 2016.

[57] See Newman 2019, ch. 2. Newton was no exception to this tendency. Recent investigations into the site of his chymical laboratory suggest he even took care to ensure the waste materials from his work would not be available to outsiders for scrutiny. See Spargo 2005, pages 319–321.

[58] Newman 2019, page 135.

early text entitled *Of natures obvious laws*, he relies heavily on the concept of a vitalist aerial niter. But by the time of his 1675 *Hypothesis of Light*, he has abandoned it.[59] Given all of this, it can be hard to understand Newton's chymical thought and what it means for his approach to short-range forces. Put differently, there is no simple or straightforward way to read his views on the nature of short-range forces off of his chymistry.

The challenges just mentioned are serious ones. But I believe there is one that is even more serious. The goal of this book is to explore Newton's *fundamental* metaphysics of substance. In the realm of material objects, this refers to the fundamental corpuscles as described by the theory of bodies articulated in *De gravitatione*. But this means there is a mismatch between the subject of much of Newton's chymical practice and the subject of this book. The reason is that much of Newton's chymical practice was focused not on the behavior of particles at the most basic level of individual corpuscles, but rather at higher-level phenomena. Specifically, Newton's view was that individual corpuscles combined with several other corpuscles to form "molecular" structures. These, in turn, would combine with other aggregates of corpuscles to form even larger entities. A very clear expression of this comes in Query 31:

> [T]he smallest Particles of Matter may cohere by the strongest Attractions, and compose bigger Particles of weaker Virtue; and many of these may cohere and compose bigger Particles whose Virtue is still weaker, and so on for diverse Successions, until the Progression end in the biggest Particles on which the Operations in Chymistry, and the Colours of natural Bodies, depend.[60]

So chymical processes are explained, in the first instance, not with reference to fundamental particles but with reference to the properties had by complex aggregate bodies. This view was not at all unusual or unique to Newton. As Newman writes:

> It is important to stress that Newton's position dovetailed closely with the much older view of many alchemists that terrestrial materials consisted of agglomerated particles (*grossae partes*) that were themselves composed of smaller particles (*subtiles partes* or *minimae partes*) that retained their full being with the larger corpuscles.[61]

[59] Newman 2019, page 442.
[60] *Opticks* 370.
[61] Newman 2019, page 180. See also pages 460–461 and Figala 2002, pages 372–373. Notably, Boyle, who exercised an important influence on Newton both generally and specifically with respect to chymistry, held a similar position. See Newman 2019, page 121 and Banchetti-Robino 2020, chs. 3 and 4.

So any effort at using chymistry to understand the nature of fundamental bodies would require the breaking up or disassembly of the larger aggregates to provide a clearer picture of the activity of particles at the lower, or more fundamental, level. It would require getting down to, to borrow a phrase from Newman, "the extreme nano-stage of corpuscular hierarchy, well below the level of gross corpuscles that made up the Lego-blocks of vulgar chymistry."[62]

To get a sense for what this means and why it might matter, we can look at a passage from the *Opticks* in which Newton seems to draw a parallel between the force of gravitation and a short-range force of attraction:

> As Gravity makes the Sea flow round the denser and weightier Parts of the Globe of the Earth, so the Attraction may make the watry Acid flow round the denser and compacter Particles of Earth for composing the Particles of Salt . . . .Now, as in the great Globe of the Earth and Sea, the densest Bodies by their Gravity sink down in Water, and always endeavor to go towards the Center of the Globe; so in the Particles of Salt, the densest Matter may always endeavor to approach the Center of the Particle: so that a Particle of Salt may be compared to a Chaos; being dense, hard, dry, and earthy in the Centre; and rare, soft, moist, and watry in the Circumference.[63]

Newton is manifestly trying to understand how short-range forces might explain the formation of salt. And he is doing so by suggesting certain particles will be drawn to the center of a mass by means of a force operating similar to the way gravity operates at the level of a planet. What is important, however, is that the entities under consideration in this passage are all still very complex. Each of them is composed of a great many other corpuscles. It seems that part of what Newton was endeavoring to do in his chymical research was break these compounds down into their more and more basic constituents. It was only by isolating the smallest possible components of matter that one might properly understand the powers had by fundamental particles, and how they behaved, when those particles were aggregated alongside one another into more complex bodies.

This, of course, represents an extraordinarily difficult challenge. In Newton's case, the challenge was made even more difficult by his experimental rigor and high standards for mathematically precise characterizations of phenomena. Newman notes that while Newton had time for some of the more "extravagant" claims of various shadowy chymical authors, he nonetheless "remained wedded to the most stringent methods of the 'experimental philosophy' and refused to believe he had succeeded at the aurific art until experiment might tell him otherwise."[64]

[62] Newman 2016, page 476.

[63] *Opticks* 361–362. For helpful background to and analysis of this passage, see Newman 2002, pages 365–366.

[64] Newman 2019, page 44.

Until he could be confident in his ability to break apart compounds and offer clear characterizations of the forces that united them, he would be loath to claim he fully understood all of the relevant powers had by fundamental corpuscles.

Ultimately, all of this means that Newton's efforts to understand and describe (let alone fully and successfully manipulate) the short-range forces governing bodies failed.[65] He was never able to offer a characterization of them that paralleled the characterization of gravitational forces in the *Principia*. In connection with this, it is telling that Newton's claim in the General Scholium that he has not found adequate experiments to characterize the powers responsible for short-range attraction was made after the period in which he undertook his most intensive chymical pursuits. Whatever other successes he might have had in coming to understand chymical compounds and their properties, I think we can take this as evidence that he saw himself as having failed to fully comprehend some of the most basic powers in nature.

6.6. The Law-Constitutive Approach

A recent and increasingly influential approach to Newton's understanding of body has been pioneered by Katherine Brading and carried forward by others.[66] On this "law-constitutive" view, part of what it is to be a body is to follow the laws of nature. More precisely, the "weak" version of the view claims that "a necessary condition for the individuality and identity of bodies is that they satisfy the laws of nature."[67] While I think this view gets some things right and points toward important truths, I also believe it is not without serious problems.[68] Most fundamentally, I think Brading is wrong to identify *laws* rather than *powers* as the constitutive components of bodies. This section examines Brading's view in greater detail in order to elucidate what I see as its shortcomings.

[65] At least, Newton's *own* attempts were failures. They did, however, provide the impetus for the important work of subsequent figures who sought to understand bodies as composites of attractive and repulsive forces. Note, however, that in this subsequent tradition it was more common to conceptualize the powers as being centered on a geometric point and not, along with Newton, as occupying a determined three-dimensional quantity of space. For excellent discussions of this subsequent tradition, see Schofield 1970 and Thackray 1970.

[66] Brading 2011 and Brading 2012. See also Brading 2013 and Brading *forthcoming*. There are hints of this kind of law-constitutive approach in earlier authors. Brading points to Stein and Janiak as interpreters offering weak or nascent forms of the view. See Brading 2012, page 24 and 30. I believe a nascent version of the view can also be found at McMullin 1978, pages 21 and 26 and at Chalmers 2009, page 132. More recently, several authors have endorsed versions of Brading's view. See, e.g., Schliesser 2011a, page 163; Smeenk and Schliesser 2013, page 143; Chen 2020. See also the distinct but, at least to my mind, related view in DiSalle 2013. Finally, it is worth noting that Robert Pasnau develops a version of the law-constitutive approach to bodies as a reading of Descartes. See Pasnau 2011, pages 316–322.

[67] Brading 2012, page 19. The "strong" version makes this both a necessary and a sufficient condition.

[68] Law-constitutivism is also critiqued in Hazelwood 2023.

The starting point for Brading's interpretation is the so-called problem of bodies, a problem Newton is alleged to have inherited from Descartes.[69] The problem arises for Descartes because he claims that extension is the essence of body. This, of course, entails a commitment to a material plenum. But Descartes often speaks as though there are multiple individual bodies, not just one indefinitely extended body. The problem of bodies asks what these bodies are. More specifically, it asks how these individual bodies are, or can be, individuated from one another. As Brading writes, for Descartes, there is "an apparent gap between what is known via the intellect (the nature of body in general, as extended), and what is known via the senses (that extended matter is divided into parts, having various shapes and motions, and that these parts correlate with our experience of bodies [plural] and their various sensory properties)."[70] The problem of bodies asks for a way to close that gap.

This problem is well-known and has received a number of detailed treatments with the result that there are various interpretations of Descartes' views.[71] A prominent strand of interpretation argues that Descartes attempts to individuate bodies through an appeal to motion. For Brading, this is suggestive of something like a law-constitutive approach. Implicit in Descartes' texts is the view that individual bodies are differentiated insofar as they are the things that are described by the laws of motion. According to Brading, Newton disagrees with Descartes on some important details, but ultimately inherits the Cartesian problem of bodies and goes much further in a clear and explicit articulation of the law-constitutive solution to this problem.

While I agree with Brading that Descartes faces difficult problems of individuation, I want to challenge the idea that this is a problem in any way inherited by Newton. My view is that Newton has a set of fundamental commitments on these issues that are radically different from those of Descartes. These radically different commitments allow him to avoid, or at least to side-step, the problem of individuating bodies. This has an important implication for the law-constitutive interpretation of Newton. In essence, it removes the motivation for holding the view or attributing it to Newton.

As we have seen, Newton was a careful reader of Descartes. And he certainly spotted something like the problem of bodies lurking in Descartes' metaphysical system. Brading's view is that he was unsatisfied with Descartes' solution and sought to offer an improved one. My view is that Newton criticized Descartes not only for his failure to solve the problem but also for endorsing fundamental commitments that generate the problem in the first place. I read Newton as claiming

[69] It should be noted that the Cartesian-oriented problem of bodies at issue in the two papers that are the focus of the discussion here is related to, but importantly distinct from and narrower than, the "Problem of Bodies" considered in Brading and Stan 2023.

[70] Brading 2012, pages 14–15.

[71] For entry points into the literature, see Sowaal 2005 and Normore 2007.

that the problems with Descartes' views on body and extension are so hopelessly intractable that the basic assumptions underlying them must be eradicated root and branch so that entirely new foundations can be laid in their place. As he writes in *De gravitatione*, he will reply to Descartes by "saying what extension and body are, and how they differ from each other . . . I consider it most important to overthrow [Cartesian philosophy] as regards extension, in order to lay truer foundations of the mechanical sciences."[72]

The primary way of doing this was to break the identification between extension and body.[73] Body is not space. Already we can see the Cartesian problem of bodies start to fade. Recall the gap that Descartes needed to close, between the truths about body delivered by the intellect (that it is pure extension) and the truths about body delivered by our experience (that it is individuated into the mobile, shaped bodies we interact with). Newton adamantly denies that we have any innate intellectual knowledge of bodies; we know about them only in virtue of our sensory experiences. For him, there is no gap to bridge.

Further, the denial of a plenum makes the identification and individuation of bodies far easier. A body, in the atomist tradition with which Newton was deeply familiar, was understood in opposition to empty space. As Charleton puts the point: "The Universe, or this adspectable World (henceforth Synonymaes) doth, in the general, consist of only two Parts, *viz.* Something and Nothing, or *Body and Inanity*."[74] This, in turn and by contrast to plenist views, offers a very easy route to the identification of bodies. Bodies are those things which are not empty space. And in the nominalist tradition with which Newton was equally familiar, individuation was something primitive. Charleton is a convenient stand-in for this tradition as well. In a passage which made a significant impression on the young Newton, and which will be discussed further in a moment, Charleton was resolute in making fundamental particles primitively individuated, speaking of "their peculiar Entity and Individuation, which is essential to them and the root of Distinction."[75] This thought was reinforced by the fact that the passage repeatedly offered the terms "atoms" and "individuals" as equivalents.[76] Newton's friend Locke is another notable exponent of this kind of approach. For him, "'tis easy to discover, what is so much enquired after, the *principium Individuationis*, and that 'tis plain is Existence it self."[77] Individuation is, again, primitive. Locke then goes on to note that if we understand an atom to be "a continued body under one immutable Superficies" (that is to say, an impenetrable solid body with a fixed size and shape) then that

[72] *DG* 35.

[73] For an account of the individuation of Newtonian bodies broadly complementary to the one developed here, see Jalobeanu 2007, sec. 3.

[74] Charleton 1654, page 16.

[75] Charleton 1654, page 108.

[76] Charleton 1654, pages 107 and 109.

[77] Locke 1975, page 330.

atom will always be differentiable from other atoms because at any given time it will be in a different place from all other atoms.[78] So individuation is, at least in theory, a simple matter.

The neat manner in which atomism and a rejection of a plenum offer a solution to the problem of bodies comes out very clearly in a quotation from one of Newton's most eccentric critics, Robert Greene, who canvased four possible views of matter.[79] He allowed that one could either endorse his view, that there were multiple different types of heterogenous matter filling a plenum, or one could choose from three other hypotheses. First, one could have the Cartesian view that homogenous matter occupied a plenum. Second, one could posit the broadly Epicurean view of homogenous particles in a vacuum; he ascribed this position to Galileo, Newton, and other "Modern Philosophers."[80] Or lastly, one could opt for heterogeneous particles in a vacuum. Notably, Greene says of this last view that it:

> has hitherto had No Patrons, nor Perhaps is likely to have Any; Because if Matter is Acknowledged to be Dissimilar in it's own Nature, there will be no Occasion for Vacuities, which are Introduced by Philosophers to Difference one Portion of Similar Matter from another.[81]

The suggestion here is that the entire point of the vacuum is to provide criteria for distinction. The Cartesian problem of bodies dissipates if one accepts atoms in the void.

Newton's facility with, and interest in, questions of identity and individuation is on display in *CPQ*. A note focused on the exposition of Genesis 1 discusses identity and individuation conditions for different species of animals.[82] Another note, "Of place," talks about individuation conditions for lines and figures.[83] Notably, Newton here uses the technical term "principium individuationis." The term would have been familiar to him from any number of Scholastic works he might have encountered as a student. And, by the time of his undergraduate studies, it had begun to find some purchase in English language prose.[84] His use of it shows

[78] Locke 1975, page 330. The claim here is not that Newton's thinking on this front is indebted to Locke. It is that Locke formulated clearly the views of a tradition with which Newton would have been familiar.

[79] Greene 1727, 934–945. Newton owned the first edition of this work. See *NL* #699. See also *NCS* #854, where Cotes discusses it in disparaging terms. Discussions of Greene frequently mention his eccentricity, philosophical and otherwise. The latter is nicely illustrated by provisions in his will "that his body should be dissected and the skeleton hung in the library of King's College, Cambridge" and for "monuments to his memory to be placed in the chapels of Clare College and King's College, in the university church, and at Tamworth." See Gascoigne 2004.

[80] Greene 1727, page 944.

[81] Greene 1727, page 945.

[82] *CPQ* 447–449.

[83] *CPQ* 351.

[84] For examples see Weemes 1632, page 30; Gillespie 1637, page 5; Jackson 1654, page 3033. Locke, as we saw above, is an example of an author using the term in English prose later in the century.

he understood that individuation presented thorny philosophical puzzles and that considerable work had been devoted to solving them.

The most important passage from *CPQ* in which Newton addresses these issues shows his familiarity and comfort with the broadly atomist and nominalist approach just sketched. It comes from Newton's discussion "Of the first matter." In this essay, much of which is drawn fairly directly from the above-mentioned passage from Charleton, Newton argues in favor of atomism.[85] Here is a key portion:

> 3. Those things which can exist being actually separate are really distinct, but such are the parts of matter. 4. Suppose the first matter one uniform mass without parts; how should that body be divided into parts, as we see it now is, without admission of a vacuum?[86]

The basic logic here is clear. We know that bodies are distinct from one another. What it takes for bodies to be distinct from one another is for them to be spatially separable from other bodies such that there could be empty space between them. And there are therefore no prospects for the successful individuation of bodies without admission of a vacuum.

This way of framing things makes good sense of Newton's "official" definitions of "body" in texts like *De gravitatione* and the *Principia*. In the former, we get that "Body is that which fills place" and place is defined with relation to space.[87] In the latter, we get that "quantity of matter is a measure of matter that arises from its density and volume jointly" and that "I mean this quantity whenever I use the term "body" or "mass" in the following pages."[88] Again, via volume, there is a clear reference to space. Arguably, what we see here is the logic of using a contrast between empty space and "something else" to offer identity conditions for bodies. Such an option is not available to the Cartesian, who is forced to equate all of extension (all of space) with body and is then very short of promising options for individuating distinct bodies from one another. To use the conceptual framework of *De gravitatione*, Newtonian bodies are determined quantities of extension with some extra feature (being full in the case of the definition from *De gravitatione* and admitting of density in the case of the definition from the *Principia*). If the considerations from the preceding paragraphs are correct, then Newton has resources for identifying and individuating bodies without recourse to the laws of nature.[89]

[85] See Charleton 1654, pages 107–109.

[86] *CPQ* 339. One might read Newton here as offering only necessary, rather than necessary and sufficient, conditions for individuation. But it is worth noting that Charleton, in the passage that is Newton's direct source here, identifies the major premise (potential separation as the hallmark of individuation) as "the general and only Rule of Distinction." Charleton 1654, page 108. Underlining added.

[87] *DG* 27.

[88] *Principia* 403–404.

[89] The approach to bodies on offer here is, broadly speaking, an atomist one. Newton's commitment to atomism receives more detailed treatment in Chapter 7, Section 7.6.

Another concern for law-constitutivism has to do with the strength of the modal claims it makes. Recall that the view claims it is a necessary condition for the individuation and identity of bodies that they satisfy the laws of nature. This indicates that the laws of nature either are conceptually prior to the bodies (i.e., there could be no bodies without the laws of nature) or it indicates that they are of equal standing such that they are mutually interdefined.[90] The concern is that, in a number of places, Newton clearly indicates that God could have made bodies that operate according to different laws of nature. And, in some of those places, he further indicates that the way that God would alter the laws of nature would be to endow the bodies with different powers and qualities than they now have.

One instance of this comes in an epistolary exchange with Roger Cotes. Cotes was charged with the revisions preparatory to the second edition of the *Principia.* Among his numerous questions to Newton was one about whether the gravitational attraction between planets and their satellites was genuinely *mutual* attraction. Is not the centripetal force, Cotes asks, more akin to a situation in which the central body stands still while the orbiting body moves toward it "by an invisible Hand."[91] As the editors of the correspondence note, "Cotes seems to have naively misunderstood the implications of the third law here."[92] It seems to me Cotes might, instead, be worried that some readers would resist the idea that gravitational attraction is genuinely mutual in instances where one object has significantly more mass than another such that the motion of the smaller body toward the larger is manifest but the motion of the larger one toward the smaller is not easily empirically verified.[93] But that is by the by. Whatever it was that was driving Cotes' worry, Newton's consideration of the question is telling.

In Newton's first reply, he offers some remarks which can be read as assuaging what I claimed were Cotes' epistemic concerns and then notes that, if there is a non-reciprocated attraction of one body to another, then "the attracted body will drive the other before it & both will go away together wth an accelerated law of motion, whereas there is no such phænomenon in all nature."[94] A draft of his response is more revealing. There he writes that:

> One may suppose that bodies may by an unknown power be perpetually accelerated & so reject the first law of motion. One may suppose that God can create a penetrable body & so reject the impenetrability of matter. But to admit of such Hypotheses in opposition to rational Propositions founded upon Phaenomena

[90] Brading 2013, page 415 is helpful on this point.
[91] *NCS* #985.
[92] *NCS* #985, note 5.
[93] For yet another take, see the discussion of this letter in Janiak 2008, pages 168–174.
[94] *NCS* #988. Cotes seems to have agreed and incorporated a similar thought into his preface for the second edition of the *Principia*. See *Principia*, 386–387.

> by Induction is to destroy all arguments taken from phaenomena by Induction & all Principles founded upon such arguments.[95]

In the first sentence, Newton affirms it as a conceptual possibility that bodies might not obey the first law of motion. A related affirmation comes in a draft of his second reply to Cotes:

> One may suppose that there may be bodies penetrable or immoveable or destitute of force, or with attraction mutually unequal, but such suppositions without any instance in Phaenomena are mere hypotheses & have no place in experimental Philosophy: & to introduce them into it would be to overthrow the Argument from Induction upon which all the general Propositions in this Philosophy are built.[96]

We certainly do not see or encounter any bodies that, for example, do not attract equally in proportion to their mass. But there is nothing incoherent or problematic in supposing that there could be such bodies. Newton's complaint is about supposing or affirming such bodies exist, not about conceptualizing such bodies. A clear consequence of this is that the laws given in the *Principia* do not determine *necessary* features of bodies.[97]

The Queries to the *Opticks* are also telling on this front. The clearest passage indicating that Newton saw bodies and their powers as importantly prior to laws is the following:

> God is able to create Particles of Matter of several Sizes and Figures, and in several Proportions to Space, and perhaps of different Densities and Forces, and thereby to vary the Laws of Nature, and make Worlds of several sorts in several Parts of the Universe.[98]

To my mind, this is a proof text. The clear sense of the passage is that laws depend on non-nomic features of bodies. It is the other, more fundamental, features

[95] *NCS* #988, note 1.

[96] *NCS* #989, note 1. See also CUL Ms. Add 3970, 620v: "Were it not for experience I should not know that matter is heavy or impenetrable or moveable. . . . And therefore to affirm anything more then I know by experience & good reasoning upon it is precarious."

[97] Indeed, these draft letters in combination with other considerations can be marshaled to argue that Newtonian bodies, strictly speaking, have no necessary or essential features at all (with the exception of being extended). See Connolly 2024. Insofar as this line of argument is successful, the attempt to argue that it is a necessary feature of bodies that they satisfy the laws of nature is misguided.

[98] *Opticks* 379–380. Underlining added. Biener and Schliesser 2017, pages 317–318 also note the relevance of this passage and the next one quoted for these issues. They also helpfully emphasize that Clarke's Latin rendering (*eoq, pacto*) is even more forceful than the English text of 1730 allows, recommending "thereby, in that manner" as a reading (footnote 25). See also Schliesser 2011b, pages 87–88.

of bodies that give rise to and determine the laws. Shortly before making this general claim, Newton had given a particular instance of it: " ... these Particles have not only a *Vis inertiæ*, accompanied with such passive Laws of Motion as naturally result from that Force ...[99] Here again it is the force that is prior to the laws that are said to result from it. The laws are, as it were, too far downstream to constitute or determine the nature of the body.[100]

It may be tempting to think that the law-constitutivist has an easy response to this sort of worry. Namely, they could argue that in the above paragraphs law-constitutivism has been mischaracterized. The reply would be to claim that law-constitutivism is *not* the view that for something to count as a body it must operate according to *the particular, contingent laws we happen to have in this spatial region and temporal era.* Instead, the view is that for something to count as a body it must operate according to *some-laws-or-other*: its behavior must be law-like. If God had made bodies that lacked inertia or operated according to an inverse-cube law they would still be bodies. It is just that we would have formulated different laws and would say that bodies qualify as bodies in virtue of operating according to those laws.

Fair enough. But, at this point, the proposal runs the risk of becoming trivial or tautological. It will hold that a necessary condition for being a body is that it behave *in a way subsumable under some law-like description.* But any behavior—even behavior that seems chaotic to finite minds like ours—is ultimately subsumable under a law-like description. In this reformulation, the law-constitutivist comes close to saying that a necessary condition for being a body is that it behave *in the way that bodies behave.* And this, of course, will be of little use unless we are already able to identify and individuate bodies.

Here is another way into the same concern. It relates to Newton's claims about his methodology. On the law-constitutivist interpretation of Newton, it is a necessary condition for something to be a body that it obeys certain laws. Something will only count as a body if it obeys certain laws. Put differently, one necessary step in correctly identifying something as a body is determining whether or not it follows those certain laws. The worry is that this implies that we must be in

[99] *Opticks* 376. Underlining added. See also a draft of this material which makes laws of motion depend on principles at CUL Ms. Add 3970.3, 620v. And see one of Newton's proposals for a revision to Book III, Proposition 6, Corollary 3 of the *Principia* at *NCS* #898, where "The force of inertia arises from [*oritur*] the quantity of matter in a body." Also striking is one of Newton's comments to Bentley: "To make this systeme therefore wth all its motions, required a Cause wch understood & compared together the quantities of matter in ye several bodies of ye Sun & Planets & ye gravitating powers resulting from thence." *NCS* #398. Underlining added. One might think the reader (Bentley) is meant to supply a suppressed "and the inverse square law" at some stage here. But, to my mind, the far more natural reading has it that the gravitational powers (and therefore the inverse square law) are portrayed as a product of the bodies.

[100] Note that these texts also cause problems for other readings of Newtonian laws, namely those which see them as "top down" impositions on bodies that literally govern them. Such an interpretation can be found in, e.g., Soto 2023.

possession of knowledge of what the laws are before we can determine whether a given thing is a body or whether it is something else (a *vody*, perhaps, which behaves exactly like a body except in very particular circumstances). But Newton's methodological pronouncements give every indication that this process works in the opposite direction. It is by observing bodies and their motions, making generalizations about those motions, and then testing those generalizations against further observations that we determine the laws. For this to work, it must be the case that we can identify and individuate bodies prior to knowing what the laws are. On the law-constitutivist view, which requires us to have knowledge of the laws in order to be sure we are dealing with bodies, it is hard to see how this Newtonian research program would be possible or how it would even get off the ground. There is a danger of getting trapped in a circle. We would need to know what the laws are in order to determine which objects are bodies. But, of course, we need to be able to pick out the bodies in order to do the natural philosophical work that would let us determine what the laws are. Indeed, it can be seen as holding us hostage. We cannot be confident we are dealing with bodies until we are in possession of the Complete Science.

Of course, it may very well be the case that this is the actual situation practicing scientists find themselves in. Brading is certainly correct that one common consequence of successful scientific research is a changing understanding and conception of the very objects that are the subject of that research. So if we are considering law-constitutivism as an abstract position in logical space, then the worry I have outlined for it may not be so serious; this is a feature of the view, not a bug. And this feature of the view might shed some very interesting light on the interplay between metaphysics and science.

My concern, however, is not about the view in the abstract. Instead, my concern is that insofar as this is a genuine feature of the view it is difficult to think that the view is one that Newton held. It sits very uncomfortably with his physical practice as well as his reflective methodological claims. Newton proceeded as though the identification of bodies was (relatively) unproblematic and the key task was to learn about their nature and behavior.

Given all this, my view is that the law-constitutive account of Newtonian bodies faces a number of difficulties. I said at the outset, however, that I think the view does capture or point toward important truths. Here, then, is one conciliatory possibility. I have said that Newton understood bodies to be composed of powers. And I have said that, for him, the task of natural philosophy is to offer precise mathematical characterizations of those powers. One way to offer such a characterization would be to articulate the correct set of laws of nature. On this view, a body's having a certain power ensures that its behavior is characterized by a certain law. Part of what it is for the body to have the power is to act in accordance with the law. Understood in this way, laws offer us a convenient mechanism for conceptualizing and communicating about the powers.

One might, then, be tempted to see my view and the law-constitutivist view as closely related or as two sides of the same coin. Ultimately, however, there are important differences which I understand as follows. First, my view is that Newton is a *power*-constitutivist about bodies rather than a *law*-constitutivist. This matters because Newton understood laws and powers to be different things. And he was clear that powers were metaphysically and explanatorily prior to laws. Second, the interpretation pioneered by Brading argues that any acceptable account of bodies requires a specification of the laws that they follow. Bodies cannot be accurately identified or individuated without recourse to the laws of nature. My version of this claim is that a *full and complete* account of bodies requires a precise description of the *powers* that constitute them. Importantly, however, I deny that such a full and complete account is needed for the identification and individuation of bodies.

6.7. Laws and Formal Causes

A view that is, in some ways, related to both Brading's and my own has been put forward by Zvi Biener and Eric Schliesser.[101] They claim that there is a fundamental and unresolvable tension in Newton's views on laws. On the one hand, Newton makes laws fundamental. Thus, they argue that Brading's view is roughly correct: laws do constitute bodies. But they also see Brading's view as incomplete. This is because, on the other hand, Newton also wants to make bodies fundamental. He allows that the laws arise from the powers and qualities of bodies. Put more precisely, their view is that "in certain passages, Newton asserts that laws are expressions of the natures of the entities they govern, and in other passages, he asserts that the natures of various entities (including forces) arise from the laws that govern them."[102]

Biener and Schliesser propose that we can better understand, and perhaps ameliorate, this tension by realizing that Newton is endorsing a form of formal causation. Formal causes, on their reconstruction, determine or constitute a substance and give rise to its features while also being posterior to a substance in the sense that they must inhere in it. Here is their characterization:

> In one sense, a formal cause constitutes (and is thus prior) to the substance in which it inheres. Without a formal cause, a substance would not be what it is or do what it does. In another sense, a substance is prior to its form. Substance is the genuine existent and subject of predication, and form's existence is grounded in the existence of substance. It is thus possible to speak both of forms as forming substances and [of] substances as grounding forms.[103]

[101] Biener and Schliesser 2017. This paper draws on some material from Schliesser 2013b.
[102] Biener and Schliesser 2017, page 317.
[103] Biener and Schliesser 2017, page 319.

Thus, on their view, Newton maintains that both bodies and laws are fundamental.

I am broadly sympathetic to the idea that we can find something like this sort of formal causation at play in Newton.[104] But I want to propose two correctives to the proposal put forward by Biener and Schliesser.[105] Taking these correctives on board will bring their suggestion in line with my own view as articulated above. The first is to insist that the formal causes of bodies are not laws, but instead are powers. I have already argued that the idea of powers constituting bodies is deep in the conceptual framework of Newton's thought. And we have just seen him claiming that laws arise from these powers.

The only serious piece of textual evidence that Biener and Schliesser cite which suggests that it is laws, rather than powers, which constitutes bodies is from Query 31 where Newton writes the following:

> These Principles I consider, not as Occult Qualities, supposed to result from the specifick Forms of Things, but as general Laws of Nature, by which the Things themselves are form'd.[106]

I freely admit that this is a confusing passage. It has caused problems for any number of commentators.[107] But one point worth making has to do with the contrast Newton hopes to bring out between his own view and a Scholastic one.[108] For the Scholastics, different species of substances (e.g., horses, oak trees, humans) had different substantial forms: hence, Newton's appeal to specific (as in *species*) forms (as in *substantial form*). Some qualities of these substances were manifest because it was clear how those qualities were connected to or flowed from the form. Others were occult because the connection between the form and the qualities was hidden or mysterious. Newton is stressing that he wants to do away with this entire apparatus in two ways. First, there is only one kind of fundamental matter, there are not heterogenous types of basic substances: hence, Newton's use of the world "general." Second, there are no substantial forms that give rise to further powers. The bodies just are the powers themselves. But terms like "power" and "quality" were familiar from a Scholastic context. Were Newton to say that the principles were qualities by which the things were formed, he would run the risk of merely further confusing the issue. By opting for "Laws of Nature," he was able to put more distance between

[104] I take the version of formal causation under discussion here to be quite different from the type attributed to Newton in Faulconer 1995 and I reject the claim of that paper that formal causation is Newton's primary or only way of conceptualizing causal processes.

[105] Biener and Schliesser's views are also critiqued in Hazelwood 2023.

[106] *Opticks* 376–377.

[107] For several examples of commentators struggling with the passage and some helpful discussion, see Hazelwood 2023, sec. 4.

[108] For more on Newton's effort to position himself with respect to Scholastic understandings of occult qualities, see Connolly 2024.

his own view and the Scholastic one while again stressing that he meant to invoke principles, that is constitutive elements, that were common to all bodies.[109]

The second revision I propose for Biener and Schliesser's view has to do with the idea that formal causes not only structure and determine what substances are and are like but also must inhere in and depend on those substances. It is the combination of these two commitments that gives rise to the puzzle and allows Biener and Schliesser to claim that, for Newton, there are two heterogenous types of *fundamentalia* in the natural world: bodies and powers (which they call laws).

On a Scholastic understanding, the formal cause of a substance is its substantial form. This is what determines a thing's nature or essence and some of the important features or properties of the substance flow from this form. The substantial form, however, is incapable of existing on its own. It must be instantiated in some parcel of prime matter. Thus, for the Scholastics, formal causation can be seen as giving rise to the sort of tension Biener and Schliesser point out: formal causes are both prior to and posterior to their substances.

My worry is that to export this understanding of formal causation wholesale into Newton's thought is to forget one of the key lessons of *De gravitatione*. In that text, Newton does allow that we should conceive of the powers constituting bodies as being something like a substantial form.[110] But he is adamant that there is no substance or substratum in which they subside. Put differently, Newton is trying to teach us that bodies might be mere forms which do not require any prime matter or substance:

> [F]or the existence of these beings it is not necessary that we suppose some unintelligible substance to exist in which as subject there may be an inherent substantial form; extension and an act of the divine will are enough. Extension takes the place of the substantial subject... and that product of the divine will is the form or formal reason of the body.[111]

There is, of course, some suggestion that space takes on the role of a substance in which the forms inhere. But, as discussed in the previous chapter, Newton then goes on to make clear that the "denomination of numerical body" is a function of the form (i.e., impressed powers). This is in marked contrast to the views of, say,

[109] In connection with this, it is notable that Newton singles out "*Quality*" along with "*Subject, Substance*, [and] *Sensible qualities*" as "Peripatetick" terms in contrast to "Mechanick ones *Body, Motions, Actions*" in a planned reprinting of the *New Theory About Light and Colours*. See the transcription of the text at Cohen 1958, page 365 and a facsimile between pages 360 and 361. On the remarkable discovery and preservation of this text, see Price 1960.

[110] Note that, on my view, the *powers* are the substantial form; I therefore think Hazelwood's reading, on which the forms in *DG* are merely spatial or geometrical entities, is too deflationary. See Hazelwood 2023, sec. 4.

[111] *DG* 43. See also page 45 where Newton talks about the effect of the divine will as the formal reason (*ratio formalis*) of bodies.

Aquinas who would have it that the quantity of prime matter (i.e., the region of space the impressed powers occupy) would be what accounts for something being denominated a numerically individual body. This means that on Newton's view "any form may be transferred through any space, and everywhere denote the same body."[112]

Therefore, my view is that a suitably modified version of the "formal causation" reading proposed by Biener and Schliesser can work. But the view will only be successful if we understand the formal causes as powers, and not as laws, and if we understand the formal causes as independently fully constituting the body, and not as needing some further entity to be predicated of or grounded in.

6.8. Mass and Density

Zvi Biener and Chris Smeenk have questioned the idea that the account of bodies presented in *De gravitatione* continued to represent Newton's favored way of thinking about key issues up through the publication of the *Principia*.[113] Their focus is not specifically on laws, but rather on alleged developments in Newton's approach to quantity of matter. More specifically, they suggest that *De gravitatione* embodies a "geometrical" approach to quantity of matter. The *Principia*, however, gives pride of place to a "dynamical" approach. But elements of the geometrical approach carry through into this later text where, Biener and Smeenk claim, they sit uncomfortably alongside the dynamic one. On the geometric understanding of quantity of matter, quantity of matter is related to volume. It is a function of the quantity of space entirely occupied by a body. According to the dynamic approach, a body's quantity of matter is a function of how it responds to impressed force.[114]

Building on a question posed by Cotes, Biener and Smeenk argue that Newton inappropriately conflates these two understandings.[115] They are, or ought to be, distinct insofar as the dynamical conception does not require quantity of matter to be invariant with respect to size. That is to say, we can imagine two bodies of equal size, one of which resists an impressed force, say, twice as much as the other. Newton, on their view, should have seen this as an upshot of the dynamical conception of mass introduced in the *Principia* but was unable to do so. Ultimately, this leaves him with a deep tension in his approach to these issues.

[112] *DG* 44.

[113] Biener and Smeenk 2012. See also the articulation of the view and role it plays in Biener 2017 and Biener 2020.

[114] For slightly different, but complementary, approaches to this dynamic conception of mass in Newton, see Janiak 2008, pages 102–112 and Joy 2012, especially sec. 3.

[115] They are followed by Schliesser 2011a, page 165 who calls the equivalence of the two conception a speculative assumption on Newton's part.

I want to suggest one way of resolving this tension. Put differently, I want to show one way in which Newton might have thought that the dynamical conception should be tethered to or restrained by the geometrical one. The first step will be to ask about the precise ontological status of the account of quantity of matter on offer in the *Principia*. The second will be to briefly look at these issues in the atomist tradition prior to Newton. The final, and most important, step will be to examine a number of Newton's own statements that bear on this point. We will see that his response to Cotes' queries actually contains a full-throated and, arguably, persuasive articulation of the view that the geometric and dynamical conceptions should be understood as aligned with one another.

Biener and Smeenk note that the "austerity of the dynamical conception of mass stems from the limited mathematical framework of the *Principia*."[116] On the one hand, this certainly gets something right. Newton is tremendously careful in the *Principia* to avoid certain metaphysical issues by offering mathematical treatments of various concepts. So, in that sense, the *Principia*'s preferred way of treating quantity of matter might be in terms of resistance to impressed force. And the resulting treatment might be relatively austere. But there is a danger of reifying this approach to quantity of matter: of saying that Newton saw this as informing us about quantity of matter's metaphysical nature. If, as seems correct, the *Principia* studiously avoids offering full metaphysical treatments of various things then this would be a mistake. Perhaps Newton's view was that these substantive metaphysical questions should be addressed elsewhere. None of this is to deny that Biener and Smeenk are correct to find a dynamical conception of mass in the *Principia*. It is just to suggest that Biener and Smeenk might be wrong in thinking that Newton saw this as a *complete* account of quantity of matter or as an account that is meant to capture deep metaphysical truths about quantity of matter.

What seems most likely to me is that Newton simply saw the geometric and dynamical conceptions as two sides of the same coin. Of course, the latter might be more useful in the context of experimental or mathematical physics. But the former might be more useful when we are trying to understand what bodies are like and when we are trying to construct a metaphysical account of corporeal substance.

Here it is worth noting that the mechanist tradition which forms an important background to Newton's thought had a certain kind of antipathy toward the idea that a body's mass or quantity of matter would not be in strict proportion to its size or volume.[117] Charleton, for example, writes: "[T]he Magnitude, or Quantity of a Body, is nothing but that kind of Extension, which amounts from the aggregate of the singular Extension of its component particles."[118] This, of course, is just a

[116] Biener and Smeenk 2012, page 116.
[117] This point is briefly noted, in connection with Biener and Smeenk, at Brading 2019, page 161.
[118] Charleton 1654, page 262.

commitment to the geometric conception of quantity of matter. The appeal of this understanding is made clear by what comes next:

> This duely perpended, no man need hereafter fear the drilling of his ears by those clamorous and confused litigations in the Schools, about the Formal reason of Quantity; for nothing can be more evident than this, that the Extension or Quantity of a thing is meerly *Modus Materiæ*, or rather, the *Matter* it self composing that thing.[119]

Charleton is channeling the reductionist appeal of mechanist philosophy. The virtue of this approach to quantity of matter is that it saves us from having to posit quantity as an additional entity in our metaphysics: a (mysterious) real quality inherent in material beings. Newton, at least in his youth, was deeply impressed by these considerations. He drew on the passage just cited as well as the discussion of condensation and rarefaction that followed it in a passage of *CPQ*, which characterizes the Scholastic view of quantity having "several proportions" to a substance, as "impossible."[120] The young Newton would have seen a similar analysis of these issues in Descartes' *Principles*.[121]

So in the tradition inherited by Newton there were clear obstacles to an adoption of the dynamical conception of quantity of matter that did not have it wedded to the geometrical one. Of course, one might respond as follows: "So much the worse for that tradition! All this shows is that Newton was visionary and insightful in his development of the dynamical conception. His predecessors were just not in a position to understand and appreciate these sorts of advances." Perhaps. But I want to suggest that Newton's replies to Cotes, in connection with his statements in other places, indicate a strong continuity in his views.

Here is the set up for Cotes' worry:

> Let us suppose two Globes *A* & *B* of equal magnitudes to be perfectly fill'd with matter without any interstices of void Space; I would ask the question whether it be impossible that God should give different *vires Intertiæ* to these Globes. I think it cannot be said that they must necessarily have the same or an equal *Vis Intertiæ*.[122]

[119] Charleton 1654, page 262.

[120] *CPQ* 339. See also *CPQ* 40–41. Isaac Barrow had also praised mechanism's ability to easily account for condensation and rarefaction along these lines. See Stewart 2001, pages 80–81. See also Henry 2024b. Brading *forthcoming* offers an alternative treatment of Newton's views on condensation and rarefaction, one in which empirical considerations predominate.

[121] *CSM*, I, 224–227. Newton references this Cartesian analysis at *DG* 52.

[122] *NCS* #893.

Cotes goes on to note that Newton consistently determines a body's quantity of matter by its inertial force but that he also consistently assumes that bodies with equal quantities of matter will occupy equal quantities of space. And this, Cotes says, Newton ought not do unless he also says it is impossible for the two equal globes to have different inertial forces. As previously mentioned, Biener and Smeenk are impressed by this reasoning. They see Cotes as right to press this point, as fully appreciating the implications of the dynamical conception of force, and as pointing toward an unjustified assumption on Newton's part. On my view, Newton's response is deeply revealing and allows us to paint a slightly different picture.

Newton's first reply is largely dismissive.[123] In essence, he just reaffirms that he sees the geometric and dynamical conceptions as being aligned with one another. One might see this as him failing to appreciate the depth of the problem Cotes had identified. Alternatively, one might see this as evidence of just how firmly embedded in Newton's mind was the idea that the two conceptions were related. Fortunately, Cotes pressed on and restated his worry.[124] Newton's reply proposes a rewording of the relevant text from the *Principia* that, in some ways, does not move very far beyond what he had already offered. Far more revealing is what Newton says about the purpose of this new phrasing. He tells Cotes that it is for "preventing the cavils of those who are ready to put two or more sorts of matter."[125] This is telling. To go back to Cotes' examples of two globes, Newton is claiming that if *A* and *B* were equal in size and both were completely solid but they nonetheless differed in their inertial forces, they would be made of different substances. Cotes' supposition is akin to the introduction of multiple different types of matter. Newton thinks anyone willing to countenance multiple types of matter is a waste of his time, a mere caviler.

Cotes is unwilling to let the matter rest here. He follows up by again zeroing in on the modal and epistemic status of the claim that all bodies have "the same Vis Intertiæ in respect of their real magnitude or extension *in Spatio pleno*."[126] He does not see how this could be known or deduced a priori as a necessary truth; indeed he points to a passage from the *Opticks* in which Newton seems to indicate it is a contingent fact. And he does not see that there is empirical evidence which establishes the geometric and dynamical conceptions as being extensionally equivalent.[127] Newton's reply rather coolly thanks Cotes for pressing the issue and

[123] *NCS* #894.
[124] *NCS* #896.
[125] *NCS* #898.
[126] *NCS* #899.
[127] McGuire also sees Cotes as offering Newton this kind of dilemma. See McGuire 1966, page 208, note 8. It is worth noting that Cotes' realization that the geometric and dynamic conceptions of quantity of matter need not align was likely one he had come to shortly before this exchange, perhaps as a result of work connected with the production of the second edition of the *Principia*. His own lectures on mechanics and hydrostatics, presumably delivered some time after he became Plumian Professor in 1707, reflect the view that the two conceptions align. The mechanics lectures include the

gives the version of the text that was ultimately printed.[128] One might think that the matter rested here. But, roughly one year later, there is another exchange between Newton and Cotes that, I believe, can profitably be read as a continuation of this discussion.

This second exchange, discussed above in Section 6.6, also starts with Cotes invoking two globes named A and B.[129] He asks about a scenario in which A remains at rest while B is moved toward it by an invisible hand. His question is about whether an observer would understand this as *mutual* attraction between bodies or just as A exerting an attractive force on B. Newton's first response asserts that the third law is empirically justified and then notes that if the attraction were not mutual then B will "drive the other before it & both will go away together wth an accelerated motion in infinitum . . . whereas there is no such phænomenon in all nature."[130] He then follows up with a letter that states the point in a more general fashion. He says that anyone who takes exception to his findings in the *Principia* " . . . must draw his objection from some experiment or phænomenon & not from a mere Hypothesis . . . "[131] A great deal of insight into Newton's thinking here is provided by drafts of these letters. For instance, in a draft of the second he writes that: "One may suppose that there may be bodies penetrable or immoveable or destitute of force, or with attraction mutually unequal, but such suppositions without any instance in Phænomena are mere hypotheses & have no place in experi[ment] al Philosophy."[132] Newton's attitude here is reminiscent of his claim that those who think bodies of equal volumes might have different quantities of matter are merely caviling. Until they can provide some positive reason for thinking bodies of equal volumes have different inertial force they are merely introducing an idle possibility, not something that requires serious consideration.

Someone sympathetic to Cotes here might return to the epistemic dilemma he offered: Is the equivalence of the geometric and dynamic conceptions necessary and known a priori or discovered empirically? Newton, I believe, does not see himself as owing Cotes an answer to this question. He will, of course, deny that

following: "Thus if an inch of Lead be six times heavier than an inch of wood, there must be six times more matter in that inch of lead than in that of wood. The wood thus has six times more pores or empty spaces. Arguing then after the same manner concerning other substances we may be furnish'd with a very good argument for a Vacuum." Bodleian Ms. Add C 272, page 5. And in the hydrostatics lectures he identifies density with specific gravity, and quantity of matter with absolute weight, and then affirms that specific gravity, quantity of matter, and magnitude are all in fixed proportions to one another. See Cotes 1738, page 62 (which corresponds to page 109 in the aforementioned manuscript). Cotes fell ill and died in the summer of 1716, not long after this exchange with Newton and presumably without making any changes to the text of his lectures. Had he lived longer and had the chance to revise them this would mark a clear instance in which we might have learned something.

128 *NCS* #903.
129 *NCS* #985. See Connolly 2024 for more detailed discussion of this second exchange.
130 *NCS* #988.
131 *NCS* #989.
132 *NCS* #989, note 1.

it is a necessary fact or one that can be known a priori. Cotes had pointed to texts from the Queries that give evidence of this. And we have just seen Newton allowing all sorts of wild metaphysical scenarios (penetrable bodies, bodies without inertial force, the third law of motion being invalid) as, at least, epistemic possibilities. And the draft letter affirms that Newton does not see "arguments a priore" as relevant.[133] But Cotes also seems right that Newton has not empirically established the equivalence of the two concepts of quantity of matter through "Induction without exception."[134] And Biener and Smeenk are correct that Newton had no empirical way to show that they were equivalent. Why, then, should we prefer to see the geometric and dynamical conceptions as aligned, rather than divergent? How can Newton escape the epistemic dilemma proposed by Cotes?

Recall again Newton's clear frustration with someone who would posit two kinds of matter where one would do. My view is that we can see this as an application of something like the first of the Rules for the Study of Natural Philosophy. Here is the rule with its explication:

> No more causes of natural things should be admitted than are both true and sufficient to explain their phenomena.
>
> As the philosophers say: Nature does nothing in vain, and more causes are in vain when fewer suffice. For nature is simple and does not indulge in the luxury of superfluous causes.[135]

Newton clearly believed that two objects equal in volume but with different quantities of matter would not be one and the same kind of thing. They would be two different kinds of matter. He also seemed to think that all of the relevant known phenomena could be accounted for on the assumption that all perfectly solid bodies of equal volume had equal mass. Seen in this way, the burden is not on Newton to defend or justify his views in light of Cotes' question. To do so, on Newton's view, would open him up to an endless series of pointless wonderings about fanciful possibilities. The burden is instead on Cotes to provide some reason for thinking nature is more complex than Newton would have it be or to point to some phenomenon which requires the positing of further causes.

Newton may have been channeling a line of thought articulated in one of More's axioms from the *Immortality of the Soul*: "Whatsoever is unknown to us, or is known but as a meerly possible, is not to move us, or determine us in any way, or make us undetermined."[136] As Newton writes to Pardies, via Oldenburg, "If one

[133] *NCS* #989, note 1.

[134] *NCS* #989, note 1.

[135] *Principia* 794. A draft conclusion to the first edition of the *Principia* contains a similar thought: "[N]ature is exceedingly simple and conformable to herself." CUL Ms. Add 4005, 25r. This text is transcribed and translated at Hall and Hall 1962, pages 320–347.

[136] More 1659, page 6. Underlining added.

argues merely from the possibility of hypotheses, I do not see by what manner he may determine anything certain in any science; since one may always contrive other hypotheses, which will seem to supply new difficulties."[137] This willingness to dismiss hypotheses or possibilities contrived at pleasure, as opposed to required or supported by some evidence, combined with a commitment to the simplicity or uniformity of nature, would be enough to set aside Cotes' worry. And there is ample evidence of Newton's commitment to the idea that there is "simplicity in all the works of the Creator."[138] One fervent and well-known statement of the view comes from an untitled treatise on Revelation:

> Truth is ever to be found in simplicity, & not in the multiplicity & confusion of things. As the world, to which the naked eye exhibits the greatest variety of objects, appears very simple in its internall constitution when surveyed by a philosophic understanding, & so much the simpler by how much the better it is understood . . . . It is the perfection of God's works that they are all done with the greatest simplicity. He is the God of order & not of confusion. And therefore . . . they that would understand the frame of the world must indeavour to reduce their knowledge to all possible simplicity.[139]

There is, of course, not a sustained argument here. But Newton's commitments are clear and serve to illuminate the line he takes in response to Cotes.

6.9. Conclusion

This chapter has looked at developments in Newton's theory of bodies after the composition of *De gravitatione*. In keeping with the considerations and chronology outlined in Chapter 4, I have tried to show that Newton remained committed to the general contours of his theory, even if he allowed that some additions to it were necessary. More specifically, Newton allowed that some phenomena were mechanically inexplicable and, therefore, thought bodies might be composed of not just powers of impenetrability, mobility, and mind-interactivity, but of others as well. One of his goals in the practice of natural philosophy was to differentiate and delineate these powers and to offer precise mathematical treatments of them.

[137] *NCS* #66.

[138] *NCS* #588. See also *NCS* #280: "Nature is after all simple, and is normally self-consistent throughout an immense variety of effects, by maintaining the same mode of operation." And see Dibner Ms. 1031 B, 4r: "Tis suitable wth infinite wisdom not to multiply causes wthout necessity." See Rogers 1999 for more on this theme in Newton's thought.

[139] Yahuda Ms. 1.1, 14r. On links between texts relating to the interpretation of Scripture, such as this one, and Newton's methodology in natural philosophy, see Mamiani 1991, pages 162–167; Mamiani 2000; Mamiani 2002, pages 396–400. See also Ducheyne 2012, pages 264–269.

This chapter has also argued that laws of nature, while important in Newton's thought, are not a fundamental element of his ontology. More precisely, Newtonian bodies are not constituted by or determined by laws of nature. Further, the chapter has argued that while quantity of matter can be understood as a dynamical concept in a way that detaches it from the amount of space a body occupies, Newton was steadfast in his rejection of the idea that perfectly solid bodies could have equal volumes but unequal masses. Further, this does not represent an oversight or lack of vision on Newton's part. Instead, it helps to shed light on some of his principled and deeply held commitments.

7

Minds, Mental Powers, and Newtonian Dualism

7.1. Introduction

Chapter 4 began with the question "What are Newtonian bodies?" By parallel, this chapter can begin with the question "What are Newtonian minds?"[1] This question will be the focus of much of the next two chapters. But we should also note, from the very outset, that when asking about the fundamental nature of Newtonian minds we are also very often asking "What is the metaphysical relationship between Newtonian minds and Newtonian bodies?" The question of *what Newtonian minds are* and the question of *how they relate to bodies* are both important and are obviously closely related. There is a danger, however, of allowing the second to eclipse the first. These chapters will, to the degree possible, attempt to keep them separate in an effort to ensure that both receive their due.

7.2. Chronological Considerations

One important reason for keeping the distinction between the two questions firmly in mind relates to the possibility of diachronic developments in Newton's thought. In previous chapters, I have argued that Newton's thinking on God remained remarkably stable throughout his career. And I have also argued that something similar is true for his thinking about bodies. He was fundamentally committed to the theory of bodies outlined in *De gravitatione*. And while the theory was enriched and details were subsequently filled in, the commitment remained constant. With respect to minds, however, I think it is possible that things are slightly different.

While Newton's views on what minds are do not appear to have greatly changed, there is a serious question of whether his views on how minds related to bodies

[1] A linguistic point: Newton very often uses the term "soul" or "spirit" in instances where I will use "mind." I think very little hangs on this terminology and mean for all three to be equivalent and capable of being used interchangeably in what follows.

Newton's Metaphysics of Substance. Patrick J. Connolly, Oxford University Press. © Patrick J. Connolly 2025.
DOI: 10.1093/9780191959240.003.0007

underwent a substantial shift during the course of his career. Ultimately, my view is that his views remained relatively stable. But the question is worth exploring for a number of reasons. First, it is important just for the purpose of giving a full and accurate account of Newton's metaphysics. Second, the question is worth looking at because it helpfully illuminates the possibility space created by Newton's account of substance and the basic metaphysical framework developed in *De gravitatione*. Third, it provides a useful context in which to explore Newton's commitment to mortalism, his relationship with John Locke, and his understanding of Locke's philosophical thought.

This chapter will examine Newton's approach to minds and to the relationship between minds and bodies in the earlier stages of his career. The next chapter will examine some later developments and consider the question of whether or not his views underwent a fundamental shift or reorientation.

7.3. The Mind in *CPQ*

This section explores Newton's early commitment to a form of substance dualism in *CPQ*. I believe his initial acceptance of the view was, in some sense, pre-reflective. It was something he simply inherited from his religious upbringing and early educational experiences. Consider the very casual division of substances into two types in one of the earliest Newtonian texts to which we have access, his "Of a Universall Language": "Now ye frame of Beings is made up of several substances (bodys & spirits) each of wch are in this or that state case or Condition."[2] The division of substances into bodies and spirits seems to slide very naturally into the sentence. Unfortunately, beyond the fact that Newton evinced a high level of comfort with the distinction, the passage can tell us little else. More specifically, the passage does not tell us what bodies and spirits are or what the nature of the distinction between them is. On the one hand, this is unsurprising given the nature and topic of the essay. On the other, my suspicion is that at this time a fairly young Newton, having simply fallen into a commitment to substance dualism, might himself have not been sure about these issues.

Part of the reason for this suspicion is the fact that the next text we have that bears on the subject, *CPQ*, shows Newton probing the value and limits of the distinction and seeking to better understand it. Put differently, it shows him moving from a pre-reflective commitment to dualism into a more considered version of the view. Mind features prominently in *CPQ* and the following passage offers a helpful introduction into one of Newton's major lines of inquiry in that text:

[2] Transcribed in Elliott 1957, page 7. Elliott argues that the text was composed either shortly before or shortly after Newton entered Trinity College in 1661.

> Philosophy ~~Occult Qualities~~ The nature of things is more securely and naturally deduced from their operations upon one another than upon our senses. And when by the former experiments we have found the nature of bodies, by the latter we may more clearly find the nature of our senses. But so long as we are ignorant of the nature of both soul and body we cannot clearly distinguish how far an act of sensation proceeds from the soul and how far from the body.[3]

There are three points to make about this passage. The first is just that it has at its heart a robust distinction between minds and bodies. This is as important as it is obvious. The second is a note of uncertainty or humility. Newton allows that the "natures" of bodies and minds are unknown to us. Third, Newton clearly believes that some phenomena, sensation is singled out here, might be attributable to bodies, to minds, or to a combination of the two.

Newton spends a lot of time in *CPQ* exploring issues connected with the fundamental nature of bodies, sometimes with reference to "their operations upon one another." But to my mind, it is unclear whether and to what extent he takes himself in that text to have found out their complete nature. Regardless, he does continue on to consider questions about the degree to which phenomena like sensation and memory are physical or mental phenomena.

This aspect of *CPQ* has received significant attention from commentators.[4] This may be, in part, because of the dramatic—one is tempted to say eye-popping—nature of some of the experiments Newton undertakes.[5] But the details of these experiments need not concern us here. Suffice it to say that Newton believes the constitution of the body, and changes it undergoes, can have a significant impact on the workings of various mental phenomena and on our phenomenological experience, especially with respect to sensation.

Recalling the two questions from the introduction above, however, we can note that these investigations are tied to the relationship between minds and bodies. But *CPQ* also contains considerations that more closely target the first question. Some of these focus on the mind's existence as something *distinct* from the body and some of them focus on the *nature* of the mind. We can examine them in turn.

Considerations relating to memory are central to Newton's claim that the mind is something distinct from the body in *CPQ*. The lengthiest treatment comes in a note entitled "Of Memory." This note shows Newton keenly interested in the way that memory could be affected by changes to physiology. Injury or illness, for example, could produce memory loss. But, importantly, it also shows him considering memory, as we might say, on its own terms. Why do we remember some things and not others? What jogs our memory? What can be remembered and what cannot? Put differently, Newton is also attempting to explore memory as an

[3] *CPQ* 377.
[4] See, e.g., *CPQ* 216–240; Iliffe 1995; Connolly 2014, sec. 7.
[5] See, e.g., *CPQ* 439–447.

ability or faculty. For current purposes, the most important part of the entry is its final sentence: "If memory be done by characters in the brain yet the soul remembers too, for she must remember those characters."[6] Newton here acknowledges that the body may well have an important role to play in memory. Like many early modern thinkers, he allows that certain experiences might leave physiological traces (characters) in the brain and that these might be partially responsible for our ability to remember those experiences. This could explain why damage to the brain sometimes affects or destroys memory. And there is also a later text which strongly suggests that the role played by the body in cases of memory might even be a necessary one.[7] But Newton affirms that whatever the physiological mechanism correlated with memory might be, it is on its own insufficient to account for the phenomenon of remembering.

We can get a sense for the argument Newton is offering by considering the difference between the following two scenarios:

> A: Looking at my planner, I realize I have a doctor's appointment in the afternoon. Knowing how immersed I can get in my day's work, I set an alarm clock to ring at the time I should leave for my appointment. The alarm clock goes off; I recall my appointment and head to the doctor.
>
> B: I see a morning weather forecast calling for rain in the afternoon. Knowing I have errands to run after lunch and wanting to remember my umbrella so as not to get wet, I lean it against the front door. Heading out later in the day, I see the umbrella and slip it into my bag.

In both cases there is some thing that serves as a reminder. But the cases are disanalogous. In A, the noise of the alarm is a reminder of the doctor's appointment. In B, the umbrella serves as a reminder *of itself*. This means that in A some further cognitive activity is needed on my part. I must recognize that the alarm is linked to my appointment and not to the kickoff time for a soccer game I want to watch on TV or the need to pick up my child from school. I take Newton to be arguing that the case of characters in the brain is more like A than like B. Whatever it is in our physiology that contributes to memory cannot itself *be* the memory.[8]

A similar line of argument is on display in a later entry, "Of the soul":

[6] *CPQ* 393.

[7] The later text is CUL Ms. Add 3970, 338v. It is a draft of a conclusion to the *Opticks* with a number of hypotheses. I.B. Cohen suggests it was composed around 1693. See Cohen 1966, page 179. The content never appears in a printed version of the text but is nonetheless revealing. "Hypoth. 5. The essential properties of bodies are not yet fully known to us. Explain this by ye cause of gravity, & by ye metaphysical power of bodies to cause sensation, imagination, & memory, and mutually to be moved by o^r thoughts." Transcribed at Cohen 1966, page 181.

[8] For a broadly similar analysis, with special attention to More's influence on Newton's treatment of memory, see Buchwald and Feingold 2013, pages 31–34.

> Memory is a faculty of the soul in some measure, for else how can diverse sounds or words excite her to diverse thoughts, or 3, or 4, or 5, or more words beget the same thought in her. Perhaps she remembers by the help of characters in the brain, but then how does she remember the signification of those characters.[9]

The second sentence reiterates the idea that a physiological mechanism could only provide a character, but something more would be needed to fix and retain the referent of that character. The point is enriched by the first sentence. Someone saying "Keegan" might cause me to remember my dog. But someone saying "dein Hund," "tu perro," "man's best friend," "the mutt that lives in your house," or any number of other things might also cause me to remember my dog. Newton is suggesting that if memory were purely physiological, words and memories would have to be correlated in a one-to-one fashion. He does not see a more complicated mechanism whereby there exist multiple routes to the same content as feasible. He may be indebted to Descartes on this front, who in the course of arguing that non-human animals were mere machines, made the following observation:

> [Non-human animals] could never use words, or put together other signs, as we do in order to declare our thoughts to others. For we can certainly conceive of a machine so constructed that it utters words, and even utters words which correspond to bodily actions causing a change in its organs (e.g. if you touch it in one spot it asks what you want of it, if you touch it in another it cries out that you are hurting it, and so on). But it is not conceivable that such a machine should produce different arrangements of words so as to give an appropriately meaningful answer to whatever is said in its presence, as the dullest of men can do.[10]

As is helpfully brought out in an article by Evan Thomas, what Descartes is emphasizing here is the way in which purely mechanical processes do not display the same *flexibility* that our conscious thought does.[11] I propose we read Newton as making a similar claim.

"Of the Soul" contains one further argument that memory cannot be a purely mechanical process. It begins with the idea that if the soul were nothing but "modified matter," then memory would consist in nothing other than "action."[12] On this view, the thought of some thing would remain in our mind as long as the action persisted and we would lose the thought of the thing when the action ceased.[13]

[9] *CPQ* 451.

[10] *CSM*, I, 140. Notably, Newton seems to have dog-eared this page in his copy of the *Discourse*. See *CPQ* 145. Descartes' views on non-human animals are also referenced at *CPQ* 421.

[11] See Thomas 2020.

[12] *CPQ* 449.

[13] Digby uses this as a criticism of Descartes' theory of memory at Digby 1644, pages 282–283. Newton owned the 1658 printing of this work: *NL* #516.

Newton's worry is as follows: "but how shall we call this thing into memory the action being done and we having no principle within us to begin such a motion again within us."[14] It is meant to be a fact obvious from introspection that we can and do conjure up memories in the mind. On a purely mechanical account there is no way to explain this. Of course, Newton has smuggled in a premise. The premise is that we can freely, or at will, bring memories to mind (rather than only have them triggered by some external stimulus). As he puts it, bringing a memory to mind must be the result of a "principle within us." This is manifestly not a premise shared by Newton's opponent here. But it is highly revealing in two ways. First, it indicates a certain approach that Newton took toward his observations of the workings of the mind. He took them to be phenomena not easily explained away. Second, it hints at an important role for the will in Newton's conception of the mind. Both of these will be significant later on.

Some other considerations from "Of the Soul" are worth mentioning in this context. Specifically, the entry contains a number of critiques of a purely mechanical process of perception. Suppose, Newton says, perception were merely a function of the animal spirits impacting a certain part of the brain, perhaps the pineal gland. Any impact would surely push that part of the brain into the animal spirits on the side opposite the initial impact. But these spirits would resist that motion and exert a new impact on that part of the brain. This means that "we should perceive things double or we should never see anything before us, but there would be some apparition behind us."[15] He also critiques a view found in Hobbes, on which motions caused by sense continue on in the brain with their original force, on the grounds that the claim fails to explain the fact that we forget things and on the grounds that it fails to capture the phenomenological distinction between sensing and imagining the same thing.[16]

These arguments, unlike those centered on memory considered above, stop short of the claim that no purely physical account of perception and imagination can be offered and therefore the mind must play some role in these processes. At the same time, their critical tenor is clear. Newton's positive view was that the mind, as something distinct from the body, does play a role in perception and imagination. This is obvious from a series of three queries that close "Of the Soul." Noting that the images of objects are projected onto the retina upside down, he asks why we do not perceive things (to borrow a charming seventeenth-century phrase used

[14] *CPQ* 449.
[15] *CPQ* 449.
[16] *CPQ* 449. *NCS* #365, a letter written to Locke in 1691, seems to offer a closely related argument targeting this Hobbesian view. Newton is there advancing the claim that the imagination can produce motions in the brain with the result of changing our sensory phenomenology. Specifically, he claimed he could, by "intending his phansy" on the image of the sun, produce an afterimage like the one he received after staring at the sun. He claims that "To place this effect in a constant motion is hard because ye Sun ought then to appeare perpetually. It seems rather to consist in a disposition of ye sensorium to move ye imagination strongly & to be easily moved both by ye imagination & by ye light."

elsewhere in the work) "with the heels upward." The answer is that: "The mind or soul cannot judge the image in the brain to be inverted, unless she perceived external things ~~immediately~~ with which she might compare their image."[17] The third query asks why, if the soul is seeing the images of things in the brain, it does not judge the things themselves to be *situated in* the brain:

> Response: Because the image of the brain is not painted there, nor is the brain perceived by the soul, it not being in motion, and probably the soul perceives no bodies but by the help of their motion. But were the brain perceived together with those images in it we should think we saw a body like the brain encompassing and comprehending, ourselves, the stars, and all other visible objects.[18]

Newton's understanding of perception here is fairly clear. The soul receives ideas on the basis of what goes on in a specific part of the brain. Our physiology is important insofar as changes to our eyes, optic nerves, or parts of the brain will change what goes on in the part of the brain to which the soul has a special relationship. So our bodies play a role in *what* we perceive but, strictly speaking, it is the soul that is responsible for perception and not the body.

Thus far we have focused on what *CPQ* says about the mind as something distinct from the body. We can now turn to what it tells us about the nature of the mind itself. True to form, Newton eschews discussion of the fundamental nature of the mind in order to focus on its observable features. Put differently, Newton's focus is on the capabilities or powers of the mind. Following standard terminology and the language Newton himself uses at one point, I will refer to these as *faculties* of the mind.[19]

From the discussion above, it is clear that sensation and memory will be two of these faculties. "Fancy," or imagination, is another faculty which receives a lot of attention in the text. Most helpful for current purposes are entries which address the relationship between imagination and other faculties. One such is the following:

> *Quaere*: Whether fantasy as well as sight is not done in a plane? Whether we can imagine two like things to be one behind another as we can do them one beside another? Resp: No, but we can understand them so.[20]

This passage shows that Newton understands imagination to be, as the etymology indicates, imagistic in nature. And, importantly, the passage indicates that we have a faculty of understanding which is distinct from imagination and which is

17 *CPQ* 451.
18 *CPQ* 451.
19 Newton refers explicitly to memory as a "faculty of the soul" at *CPQ* 451.
20 *CPQ* 387.

non-imagistic.[21] Finally, while it is not mentioned explicitly, Newton also believes there is a faculty of the will. One of the arguments about memory discussed above showed that Newton thought we had (at least some) voluntary control over what we brought to mind. And a number of experiments Newton conducted with the afterimage created from staring at the sun showed he believed in our volitional control over the imagination.[22] With enough effort, Newton concluded, one can even cause the imagination to produce changes in the optic nerve.

McGuire and Tamny have argued that *CPQ* contains a tension between competing strands of Newton's thought regarding the mind. They identify, on the one hand, Hobbesian theses to which Newton was attracted. Specifically, they claim that Newton shows sympathy for Hobbes' view that some cognitive tasks could be achieved by the body alone and his view that all thought is derived from sense-experience and is therefore imagistic. On the other hand, there is a line of thought that runs counter to Hobbesian imagism and which holds that there are "various ways of knowing specific to the mind's cognitive powers."[23] They attribute this to Descartes' influence. But, on their view, this is a "more subtle" component of *CPQ* and Newton never explicitly distinguishes between the senses, the imagination, and the understanding.

On the basis of the discussion above, I think McGuire and Tamny's analysis should be rejected. Newton was passionately interested in the way the body contributed to the content of perceptual experiences. But he never seemed to take seriously the idea that the body could be *fully* responsible for any of our perceptual or cognitive activity.[24] So I think it goes too far to claim that Newton was "impressed by Hobbes's claim that cogitation, sensation, imagination, and perception are nothing but motions."[25] And the text *does* seem to explicitly advocate for a distinction between different faculties. At a minimum, it contains an explicit endorsement of non-imagistic cognitive content. We can understand one object being situated behind another even though we cannot perceive or "image" the scenario. Put differently, there is no tension in Newton's thought. He rejects the first strand of thought and accepts the later. In addition, it is not entirely clear that the second strand of thought should be classified as Cartesian. The thought-experiment Newton offers

[21] Newton's note "Of Quantity" at *CPQ* 453 also gestures at a distinction between imagination and understanding in a way that foreshadows his clearer treatment of these faculties as they relate to extension in *De gravitatione*. For some discussion, see Iliffe 2017, pages 96–97.

[22] *CPQ* 397 and 443–447. The experiment is also briefly mentioned in "Of Colours." CUL Ms. Add 3975, 16. Newton reports on these experiments in a 1691 letter to Locke. See *NCS* #395. His descriptions and analysis align neatly with the text of *CPQ* suggesting continuity in his views. Strikingly, Newton seemed to recall these experiments, and perhaps still suffer their aftereffects, even up to the time of his death. See Keynes Ms. 130.15, 1.

[23] *CPQ* 321. Though see also their claims at *CPQ* 217–218, which are much more amenable to my interpretation.

[24] In this sense, my analysis is closer to that offered in Iliffe 1995, sec. 2; Dempsey 2006, page 421; Buchwald and Feingold 2013, page 32.

[25] *CPQ* 133.

with one object behind another seems structurally disanalogous to the Cartesian examples of the sun and the chiliagon. And while Descartes was a prominent advocate of the distinction between the senses and the understanding, non-imagistic theories of cognition were also standard fare in the Scholastic philosophy of mind that Newton might have encountered. Admittedly, some of Newton's thinking on the topic does seem to have been driven by issues of the infinite and the indefinite which he encountered in Descartes.[26] What seems correct to me, however, is that while Newton may have been influenced by others on this front, he developed and endorsed the view within his own conceptual framework.

7.4. Minds in *De gravitatione*

Commentators have taken different approaches to the ontology of minds in *De gravitatione*. Kochiras argues that it shows Newton "vigorously" defending dualism.[27] In sharp contrast, Dempsey has claimed that it shows Newton adopting a form of monism.[28] And others have claimed that the text shows Newton had little interest in fundamental questions about the ontology of minds, choosing to focus instead on their behavior.[29] This section will argue that Newton continued to endorse a dualist position in *De gravitatione*. But, as suggested above and as will be discussed in the next chapter, *De gravitatione* also could have laid the groundwork for Newton to see beyond this position and consider a form of agnosticism that better conforms to his core commitments.

We can start by considering Liam Dempsey's claim that Newton endorses a form of monism in *De gravitatione*.[30] Dempsey's strategy for arguing this point has a philosophical component and a textual component. The philosophical component relies on the claim that Newtonian minds and bodies are fundamentally ontologically interdependent. The argument stems from Newton's criticism of the Cartesian claim that body and mind are heterogeneous. Newton, like so many others, is worried that if mind and body are *fundamentally* different then they will

[26] On this point, see McGuire 1983.

[27] Kochiras 2013, page 328.

[28] Dempsey 2006, sec. 2 and Dempsey 2009, sec. 4.

[29] Stein 2002, page 282 and Slowik 2013, page 431.

[30] This may represent a slight mischaracterization of Dempsey's views, which are somewhat complex, as the following quotation indicates: "I have argued elsewhere that, with respect to the mind-body relation, Newton defends a kind of substance monism, at least in one important sense of the word 'substance'. I have also advanced the view that [some of Newton's views] imply a sort of idealism. Here I wish to emphasize the aspects of his view that are dualistic in nature, aspects that follow from his idiosyncratic idealism." Dempsey 2017, page 459. I find this view to be somewhat enigmatic and have worries about its stability. For the moment, I will confine my analysis to the monist elements of Dempsey's view. I focus on the idealist and dualist elements elsewhere. Schliesser also attributes a form of mind–body monism (a property dualism on which sentience is an emergent feature of matter) to Newton and tentatively endorses some of Dempsey's claims. See Schliesser 2021, pages 63–67.

not be able to interact. But the interactivity of minds and bodies is central to what they are as substances. Any account of mind and body which makes them unable to interact misses something essential about them as substances.

So far so good. Newton does end up constructing an account of bodies that makes the ability to interact with minds one of their constitutive powers. And he continuously emphasizes that our minds can exert causal pressure on bodies and the ways in which this makes us similar to God. So part of what it is to be a body is to be capable of standing in certain relations to minds. And part of what it is to be a mind is to be capable of standing in certain relations to bodies. Dempsey invokes a Cartesian conception of "substance" which holds that something is a substance if it depends on nothing other than God for its existence. And he then infers that these facts about bodies and minds make them mutually dependent and therefore bodies and minds do not exist as distinct substances.[31] Thus, Newtonian monism.[32]

The problem is that the facts pointed out by Dempsey will, at most, only establish a kind of *conceptual* dependence between minds and bodies and not the *ontological* dependence his argument requires. Put differently, these considerations will not show that minds and bodies are incapable of existing apart from one another. The reason is that something may have a certain disposition without that disposition ever being manifested. Magnets have the power to attract iron filings. This does not make magnets ontologically dependent on iron filings or iron filings ontologically dependent on magnets. In a world containing only a magnet and no iron filings, the magnet would still have the power or disposition to attract iron filings *were any to be present.* And a small magnet might come into being and pass out of being in a world that contains iron filings without ever coming near enough to trigger its disposition to attract. Similarly, the mere fact that minds and bodies are disposed to interact with one another does not entail that they must of necessity do so.

This is especially easy to see in the case of bodies. Newton makes clear in *De gravitatione* that mobility and impenetrability will enable all bodies to interact with human sense organs but that this will be insufficient for causing perceptions in mind. Interaction with the mind is a power only exercised by bodies actively united to minds in a special part of the brain. But just because other bodies are not exercising the power does not mean they lack it. Newton reasons as follows:

31 Dempsey 2006; Dempsey 2009, sec. 4; Dempsey 2011; Dempsey 2014.

32 Dempsey allows that this form of monism resists "neat characterization." Dempsey 2011, page 242. It might be noted, however, that in many ways, the view Dempsey ascribes to Newton (and sometimes Locke) could arguably be characterized as hylomorphism. Just as a body and substantial form have "distinct characteristics and functions" the two are "nevertheless, *ontologically* interdependent" in that they can only exist united together as a hylomorphic compound. Dempsey 2011, page 242. See also Dempsey 2006, page 425: "For Newton, then, mind and body do *not* possess independent existence . . . . Human minds exist only in living human bodies, and human bodies owe their life and motion to the minds with which they are coupled." While Newton does sometimes reference hylomorphism in explaining his metaphysical views, I know of no passage where he uses it to explicate the mind-body relation.

> But should anyone object that bodies not united to minds cannot directly arouse perceptions in minds, and that since there are bodies not united to minds, it follows that this power is not essential to them, it should be noticed that there is no suggestion here of an actual union, but only of a capacity of bodies by which they are capable of such a union through the forces of nature. From the fact that the parts of the brain, especially the more subtle ones to which the mind is united, are in a continual flux, new ones succeeding those which fly away, it is manifest that that capacity is in all bodies.[33]

Many bodies currently possess the capacity but do not exercise it.[34] And it is possible to imagine a body which, for whatever reason, *never* unites with a mind. It will still be true that it exists and it will exist without ontologically depending on any minds.[35] So *inter*dependence between the substances is ruled out.

What about an asymmetric dependence of minds on bodies? The same basic philosophical point should apply. It is entirely possible that there is a thinking substance with the disposition to unite with bodies, but this will not show that it ontologically depends on bodies. So I think we ought to reject this part of Dempsey's view as well; it is simply not clear that Newtonian consciousness is a *necessarily* embodied phenomenon (even if natural or divine processes always ensure that, for humans, it happens to be embodied). Two considerations relating to God should also help to establish this. First, there is no reason to think that Newton's God lacks consciousness or conscious awareness. Newton's insistence on our similarity to God and the various sensorium passages suggest that there is at least some divine analog of consciousness. But God is not embodied. So, at least in this instance, consciousness or mindedness and union with a body come apart. Second, there are considerations relating to the divine will. Could God create a mind with conscious thought not connected to a body? Or could God separate a mind from a body while allowing it to retain its consciousness? It is very hard to see why the answer to either of these questions would be no. And a manuscript passage that comes under the heading "Of God" gives some positive reason to think the answer is yes: "what ever I can conceive without a contradiction, either is or may bee made by something that is: I can conceive all my owne powers (knowledge, activating matter etc) without assigning them any limits Therefore such powers either are or

[33] *DG* 48–49.

[34] Dempsey himself offers a similar analysis of this passage. See Dempsey 2009, page 48 and Dempsey 2014, pages 95–97.

[35] Sometimes Dempsey's claim seems to be something slightly different. Specifically, the claim sometimes seems to be not that bodies *simpliciter* ontologically depend on minds but that *living biological human* bodies ontologically depend on human minds. This claim is plausible, but runs the risk of triviality. A dualist would assent to this claim no less than a proponent of the monism Dempsey outlines. So it is hard to see how the claim could bear much argumentative weight.

may be made to bee."[36] So minds do not seem to bear any necessary ontological dependence on bodies.[37]

Before moving on to consider the textual evidence Dempsey presents for his interpretation, we can note two other theoretical challenges it faces. The first stems from the material in the block quotation above. Newton here specifies that there is no connection between the mind and the *entirety* of the human body. The mind only interacts with a specific part of the brain. Further, the corpuscles that comprise that part of the brain are in constant flux. This raises some puzzles for the monist claim that the mind and body form one substance. The mind is not united to the entirety of the human body. And the part of the body it unites to is clearly an aggregate. But it is not a stable aggregate; its constituent parts constantly shift. In what sense, then, should we see the mind and (diachronically distinct) parts of the body as one single substance?

The second theoretical challenge is related. Newton clearly sees the power to interact with minds as one of the constituent powers of bodies. Bodies are able to "excite various perceptions of the senses and the imagination in created minds, and conversely be moved by them."[38] He is particularly forceful on this point when contrasting his view of bodies with the Cartesian view which makes them merely extended substances. He insists that his bodies are different because they also possess "that faculty or power by which they stimulate the perceptions of thinking things."[39] The worry is that on a monist account, where the mind and body are one and the same thing, it becomes difficult to make sense of this power or the need for this power. If there is only one substance, the human being, why should it need a power to interact with itself?

This brings us to the textual component of Dempsey's strategy. There is one passage from *De gravitatione* that is central to Dempsey's monist interpretation and which is deployed in a number of places.[40] It reads as follows: "[T]he distinction between these ideas [extension and thinking] will not be such that both may fit the same created substance, that is, but that a body may think, and a thinking being be extended."[41] The first thing to note is that, even taken at face value, this passage is insufficient to do the work Dempsey assigns it. All the passage would seem to

[36] Dibner Ms. 1031 B, 4v.

[37] Though they may bear, as noted earlier, a conceptual dependence in that one of their defining features might be the ability to interact with minds. And they might also bear a *functional* dependence on bodies in that they require union with a body for the execution of certain functions such as perception or volition.

[38] *DG* 43.

[39] *DG* 48.

[40] The passage plays a key argumentative role at Dempsey 2006, page 425; Dempsey 2009, page 49; Dempsey 2011, page 254; Dempsey 2014, page 110. Guicciardini 2018, page 134, also cites the passage in support of a conclusion similar to Dempsey's.

[41] *DG* 46. Dempsey sometimes offers a plausible amendment to the translation, using "the distinction between these ideas will not be so great" rather than "the distinction between these ideas will not be such that" to capture "distinctio Idearum non tanta erit." On this point, see Gorham 2011b, note 38.

allow for is the *possibility* that a single substance could be both mind and body. But Dempsey relies on the claim to show that bodies and minds are either *likely* co-instantiated in a single substance, are *in fact* co-instantiated, or are even *necessarily* co-instantiated. These readings surely go beyond what is licensed by the text.

More importantly, we should wonder whether this passage even shows Newton committing to the *possibility* of monism. Here we must look more carefully at the context of the passage and what it is doing in the larger argumentative dialectic of *De gravitatione*. The passage occurs in a paragraph that begins with the following:

> [T]he usefulness of the idea of body that I have described is brought out by the fact that it clearly involves the principal truths of metaphysics and thoroughly confirms and explains them. For we cannot posit bodies of this type without at the same time positing . . . that they are beings distinct from created minds, but able to be united with minds.[42]

An immediate worry for Dempsey's interpretation is that Newton here seems to make a distinction in kind between minds and bodies a principal truth of metaphysics. Newton then moves to illustrate the comparative superiority of his views to those of Descartes. He does this, as so often in *De gravitatione*, by pointing to Descartes' identification of extension with body. Newton charges that this means minds will have to "exist nowhere" rending their interaction with bodies (even more) mysterious. He then offers Descartes a dilemma:

> Moreover, [A] if the distinction of substances between thinking and extended is legitimate and complete, [B] God does not eminently contain extension within himself and [C] therefore cannot create it; but God and extension would be two separate, complete, absolute substances, and in the same sense. But on the contrary [D] if extension is eminently contained in God . . . certainly the idea of extension will be eminently contained within the idea of thinking, and hence [E] the distinction between these ideas will not be such that both may fit the same created substance, that is, but that a body may think, and a thinking being be extended.[43]

Having, to his mind, vanquished the Cartesian, Newton immediately continues the project of the paragraph by showing how his views are also superior to the "common idea" of bodies.

What, then, is going on in the crucial passage? As suggested above, Newton is offering Descartes a dilemma. Descartes is committed to [A] the idea that

[42] *DG* 45.
[43] *DG* 46.

extension and thought are mutually exclusive concepts. If this is true then either [B] God does not eminently contain extension or [D] God does. The first horn of the dilemma, [B], is unattractive because it entails [C]: God cannot create bodies. The second horn of the dilemma, [D], is unattractive because it entails [E]: the possibility that one substance is both extended and thinking. But the point of the second horn is to operate as a *reductio* on [A]. That is where this part of the argument gets its bite.

Given this reconstruction, we are now in a position to see what goes wrong with attempts to read Newton's views on the ontology of minds and bodies off of this passage. Simply put, Newton is not here speaking in his own voice. He is stepping into the Cartesian framework and illustrating what he takes to be problems internal to that framework. The context is Newton's effort to show that his own views are superior and so it is important the he does not see himself as facing these problems. Why not? Because he rejects [A]. For him, the distinction in substances between extended and thinking is neither legitimate nor complete. This is because he believes all substances must, of necessity, be extended. So being extended cannot be a criterion used to differentiate between different types of substances. Seen in context, then, the passage Dempsey relies on for his interpretation should not be taken as a statement of Newton's positive views and can do very little to establish that Newton adhered to a form of monism.[44]

So we have seen that Newton clearly rejects monism in *De gravitatione*. What about the claim that Newton seeks to set aside questions about the metaphysics of mind in that text? Slowik, for example, has suggested that Newton attempts "to blur, or minimize, any sharp distinction between mind and body."[45] Stein offers a related but slightly different take. According to him Newton's "suggestion sets aside the distinction of 'kinds of substance': mind-body dualism or monism, in favor of the program: *to seek to understand mental attributes and their relation to corporeal ones*."[46] The picture here is that Newton does not really seek to engage with, let alone settle, fundamental questions about the ontology of mind in *De gravitatione*.

Slowik's suggestion is correct in one sense. While I think it goes too far to claim that Newton wants to blur the distinction between mind and body, it is true that he wants to avoid certain kinds of sharp distinctions. More specifically, as we saw above, Newton is opposed to a sharp Cartesian distinction between mind and body that makes one spatial and the other non-spatial. Whatever differences exist between them, they must both be capable of location in space. Stein's suggestion also contains an element of truth. Newton is passionately interested in the

[44] A very similar analysis is offered in Kochiras 2013, pages 338–340.
[45] Slowik 2013, page 431.
[46] Stein 2002, page 282. Emphasis in original.

relationship between minds and bodies. And, further, he also offers an account of minds and bodies that focuses on their attributes, or perhaps better, their operations. This is at least true of bodies; they just *are* the powers that produce their characteristic actions.

But while there is much to recommend the view taken by Slowik and Stein, it still seems that Newton is committed to a form of substance dualism in *De gravitatione*. After all, Newton does make the distinction between bodies and minds one of the "principal truths of metaphysics" and further, one of the truths that his account of bodies is meant to involve, confirm, and explain.

This means that Kochiras' reading is closest to correct; *De gravitatione* contains a commitment to and articulation of a certain form of dualism. But, even here, I think we are in a position to offer a corrective. Kochiras claims that *De gravitatione* contains a "spirited defense" of dualism. Insofar as Newton allows that the distinction between minds and bodies is a principal truth of metaphysics something like this is correct. But, Newton's spirited *defense* is just that, a defense. There is, as far as I can tell, nothing in *De gravitatione* that constitutes a positive argument in favor of dualism. Newton might be better read as engaging in an internecine dualist dispute; he is diagnosing some problems with Descartes' understanding of the view and offering a superior way forward. Thus, I am inclined to read *De gravitatione* along the same lines as *CPQ*. It is an attempt to articulate and better understand a dualism that Newton had simply accepted as a default position.

We should now turn to what *De gravitatione* says about the nature of mind itself. There is not a great deal that bears directly on the issue in this text. But the claims Newton does make are broadly consistent with themes found in *CPQ*. What Newton says about mind–body interaction shows that he continued to be interested in both perception and the will as powers of the mind. More intriguingly, the text shows Newton reaffirming his commitment to a distinction between imagination and understanding:

> If anyone now objects that we cannot imagine extension to be infinite, I agree. But at the same time I contend that we can understand it. We can imagine a greater extension, and then a greater one, but we understand that there exists a greater extension than any we can imagine. And here, incidentally, the faculty of understanding is clearly distinguished from imagination.[47]

Unlike in *CPQ*, Newton does not mention memory explicitly in *De gravitatione*. But surely this is a result of the subject matter of the essay and not the result of Newton abandoning it as a mental faculty.

[47] *DG* 38.

7.5. The Ontology of Minds

What, for Newton, are minds? A good first step in answering this question involves a return to Newton's deflationary account of secondary substances. Minds are secondary substances and therefore have no substance at their core. What, then, constitutes them? I propose that they just are their powers. On the one hand, this is a slightly speculative proposal. There are, unfortunately, only so many texts where Newton addresses the issue directly. This makes it difficult to have complete confidence in the proposal.[48] On the other hand, the proposal seems to be on solid footing given what we know about Newton's ontology of secondary substances and the options open to him.

So my view is that the Newtonian account of human minds has them consist of spatially situated and unified powers of perception, imagination, understanding, memory, and volition. Just as God creates bodies by situating three powers in space, he creates minds by situating five powers in space. Recall that, as secondary substances, human minds cannot have a substratum or substance at their core. This means that if minds are not to be collections of powers, as I am suggesting, then there will be very few other options. One would be to posit a single multitrack power that is responsible for perception, memory, understanding, and the rest. Another would be to select one power to be primary and allow that the others flow from or are consequent on it. Neither of these seem very promising. This is for two reasons. First, both would be open to the charge of making the "primary" power act as a substratum. And this is not on the table for Newton. Second, as discussed above, Newton is impressed by the heterogeneity of our mental powers. This is perhaps most clear in the case of imagination and understanding. But it also comes through in his analysis of volition and perception. Given this heterogeneity, the prospects for amalgamating the powers together or reducing some of them to another of them seem limited. And these considerations, combined with the fact that Newton had already constructed a parallel theory of material substances as composites of powers, ground the interpretation I have offered.

Recall from Chapter 5 that Newtonian bodies faced the "glue problem" of explaining how the three heterogeneous powers are united. I argued that the power of mobility could serve as the glue uniting the powers. What is the glue for Newtonian minds? Simply put, I worry Newton does not say enough to offer a clear answer to this question. But here is a tentative proposal.

[48] On this point, I have a certain sympathy with Peter Anstey's reconstruction of some of Robert Boyle's views on the mind. He notes that—given the relative paucity of texts—there is a danger of imposing too much systematicity on Boyle and attributing to him views he himself never consciously worked out or held. But Anstey justifies his reconstruction, in part, by noting that "the answers to these questions are extremely enlightening for the modern commentator who is attempting to understand the contours of his [Boyle's] thought and his place within his broader intellectual milieu." Anstey 2000, page 187.

The proposal relates to one important feature of the ontology of Newtonian minds not yet mentioned. Specifically, it was immensely important to Newton that humans, as "holy writ testifies," are made in the image of God.[49] Among other things, this means that our minds are "of a far more noble nature than body."[50] Recall that Newton's God is a God of dominion and that the omnipresent divine will can control all aspects of creation. Newton, while always sensitive to the general nature of the *imago dei* teaching, does sometimes lay special emphasis on the similarities between our wills and God's will. And, in this sense, I think it is possible that the will may be what is responsible for "controlling" the other faculties in the sense of maintaining and coordinating their union.

To be clear, the suggestion is not that we must, at every moment, consciously and deliberately will to keep our minds united lest they disintegrate. Fortunately, or perhaps unfortunately, we can no more choose to detach our memory from our other faculties than we can choose to move objects without touching them. Instead, the reason for positing the will as the glue relates to the governing and coordinating role it plays within the mind's ecosystem.[51] More specifically, the will can—within limits—modify or modulate the other powers. It can make perception focus on some things or attend to other things. As noted above, it can get the memory to bring certain things to mind. It can direct the imagination. This is a limited kind of control, but it is control. And it points toward a binding or organizing kind of function; the will can coordinate the other faculties.

7.6. Characterizing Newton's Dualism

What then is Newton's position? Thus far, I have claimed that the text shows Newton's commitment to a form of dualism. But this is a very general claim. Situating Newton with respect to some other views in the period will help to clarify his view. Notably, Newton's dualism is not *Cartesian* in nature. Newton was certainly familiar with Descartes' views on the issue; in the Trinity Notebook he had carefully compiled a list of eight passages in which Descartes discusses the "distinction between the mind and the body."[52] But even at this early stage, issues relating to space differentiate the two thinkers. His theorizing about bodies and the vacuum in *CPQ* shows he firmly rejected the Cartesian equation of body with extension. And an entry, "Of Motion," has Newton making the following claim: "It is true God is as far as vacuum extends, but he, being a spirit and penetrating all matter, can be no obstacle to the motion of matter; no more than if nothing were in

[49] *DG* 44. On this point, see also Connolly 2014, sec. 9.
[50] *DG* 44.
[51] My thinking on this point is indebted to the account of non-agentive power and agentive abilities developed in Perler 2024.
[52] *CPQ* 465.

its way."[53] While nothing in *CPQ* explicitly affirms the spatial location or extension of human minds, the text of *De gravitatione* does. So, at a very early stage, Newton had already rejected the fundamental basis of Descartes' dualism.[54]

An alternative take on this issue has come from Gorham. Gorham emphasizes *divisibility* rather than mere *spatiality* as the key aspect of Cartesian dualism. This allows him to claim that Newton's dualism is, in fact, indebted to Descartes and falls into a distinctly Cartesian vein. While both minds and bodies are spatial on this view, they have different ways of being extended. Bodies are divisible in the sense that they are composed of *real* parts: parts which are really distinct and are separable from one another. Minds, on the other hand, are extended but lack real parts.[55] This is an intriguing suggestion, both as regards what is most central to Descartes' distinction between mind and body and as regards Newton's position. For current purposes, we can set aside the issues of Descartes interpretation and just focus on what Newton thinks about parthood and divisibility.

Early modern thinkers adopted divergent and conflicting approaches to mereological issues. Before exploring and evaluating the interpretation of Gorham on these issues, it will be helpful to introduce a number of distinctions and fix some terminology. The basic framework I will be adopting is the one developed in Thomas Holden's excellent book *The Architecture of Matter*.[56] Some objects are divisible into parts through natural processes. I can disassemble a castle built of Lego blocks. I can cut a block of cheese into two. The Lego castle and the block of cheese are therefore *physically* divisible, they have *physical* parts.[57] Whether something is physically divisible is, in some sense, an empirical question. Suppose, however, we discovered bodies that were physically indivisible. There is still a question about whether they could be split. Suppose God, for example, wanted to divide such a body in half. If such a division were possible by means of omnipotence, then it would be the case that the body was *metaphysically* divisible, it would have *metaphysical* parts. Suppose, for whatever reason, God could not divide a physically indivisible body. Would such a metaphysical indivisible body be utterly partless? Not necessarily. If the object were extended, it would still be possible to differentiate its spatial parts. For example, we could differentiate the left half of the body

[53] *CPQ* 409.

[54] For more on Newton's rejection of Cartesian dualism understood in terms of extended and unextended substances, see Dempsey 2006, sec. 2 and Janiak 2008, page 142.

[55] Gorham 2011b, sec. 3. See also Gorham 2011a, footnote 78. Note, however, that Gorham's interpretation denies that bodies are, strictly speaking, substances. Gorham 2011b, page 27. So this will be a dualism, but not a substance dualism and in that respect somewhat different from Descartes' dualism.

[56] Holden 2004. Note, however, that with respect to Newton's views the interpretation developed here departs from Holden's.

[57] As the examples are meant to suggest, I am setting aside, for the moment, the question of whether the parts the objects have are actual or potential parts. It is natural to think of the Lego castle as having actual parts (the "fundamental" Lego blocks composing it) and the cheese (a homogenous mass) as having potential parts.

from the right half. So even if a body were metaphysically indivisible, it would still make sense to talk about the body as *formally* divisible or as having *formal* parts.[58]

We can start by asking whether Newton thought bodies had physical parts. Here, it is important to distinguish between macro-bodies (my office chair, a loaf of bread, etc.) and fundamental corpuscles. The former are obviously physically divisible; our question is about the latter. As Janiak has pointed out, Newton was keenly aware of the distinction between formal divisibility, on the one hand, and physical divisibility, on the other.[59] And he rejected the view that the former straightforwardly entailed the latter. Newton was familiar with the work of Henry More, in particular More's *Immortality of the Soul.* And More had developed the concept of *indiscerpibility* to describe entities which were extended (and thus formally divisible) but which were nonetheless incapable of being broken into parts by natural forces. No doubt reflecting some Morean influence, Newton's earliest treatment of fundamental particles in *CPQ* made it clear that he wanted to articulate a theory that made them indivisible in the sense of not divisible into physical parts.[60] Janiak argues that in public-facing texts like the *Principia* and *Opticks*, Newton adopts a more epistemically humble view and argues that the question of whether matter is infinitely divisible or whether it is instead composed of indivisible fundamental atoms is an empirical one that we lack the evidence to settle.[61]

This claim of Janiak's is plausible. And Newton certainly had a general tendency toward metaphysical caution in his public pronouncements. But even in published texts like the *Opticks* and *Principia*, there are passages where Newton can be read as subtly showing his hand in a way that makes clear his opposition to the infinite divisibility of matter. In the *Opticks*, for example, he writes that in the beginning God made particles " ... so very hard, as never to wear or break in pieces: no ordinary Power being able to divide what God himself made one in the first Creation."[62] As for the *Principia*, we can pick up on an interesting point raised by Mary Domski in connection with the Third Rule.[63] In his explanation of the rule, Newton notes that

[58] A helpful introduction to these concepts and more formal characterizations of them can be found at Holden 2004, pages 9–16. Holden also introduces *intellectual* divisibility. Something has intellectual parts if a mind can represent it as containing diverse parts, even though those parts are not physically or metaphysically separable and cannot be understood as occupying distinct parts of space.

[59] See Janiak 2000. Janiak uses "mathematical" to pick out what I am calling formal divisibility. Palmerino 2013 helpfully situates Newton's view of the distinction with respect to others in the period.

[60] On this point, see Reid 2022, pages 1494–1496, which also notes the influence of Charleton. Reid ends up agreeing with Janiak that the mature Newton took atomism to be an empirical issue.

[61] Another view which makes Newton's approach to these issues subject to empirical findings is found at Brading 2011, pages 143–145. See also Brading 2013, sec. 5.3.

[62] *Opticks* 376. The passage continues on in a similar vein and references "permanent Particles." Note that this quotation comes from the Queries. Perhaps this is further evidence for Janiak's claim. Newton is advancing this positive metaphysical view in a part of the book where he has given himself more latitude for speculation. But in texts where he is holding himself to more rigorous evidential standards—the body of the *Opticks* or the *Principia*—he adopts the more epistemically humble view. Newton himself hints at something like this at CUL Ms. Add 3968, 414v, 571v, and 588r.

[63] Domski 2021, pages 42–43 and 64–66. Note that while I mean to pick up on the puzzle raised by Domski, I draw a conclusion different from hers.

divided parts of bodies can be separated from one another and that the undivided parts are at least formally distinct from one another. He then says the following:

> But it is uncertain whether those parts which have been distinguished in this way and not yet divided can actually be divided and separated from one another by the forces of nature. But if it were established by even a single experiment that in the breaking of a hard and solid body, any undivided particle underwent division, we should conclude by the force of this third rule not only that divided parts are separable but also that undivided parts can be divided indefinitely.[64]

What should stand out in this passage is that Newton thinks we are entitled to infer something on the basis of *even a single experiment*. This should stand out because Rule Three is meant to be about what pertains to "all bodies on which experiments can be made" and we are meant to be uncovering the "qualities that square with experiments universally."[65] Why, given the way the rule is set up, should a single experiment allow us to infer something about all bodies? One thing that might make sense of this would be to allow that Newton was antecedently committed to the physical indivisibility of basic corpuscles. If the default assumption was that matter was not infinitely divisible, then a single experiment would be probative and would require the assumption to be abandoned. So I take these two passages to provide good evidence that Newton believed fundamental bodies were at least naturally indivisible.

Are Newtonian bodies metaphysically divisible? Here again, I think the answer is negative. But seeing why this is so requires us to resist what initially seems like a compelling argument along the following lines. We can clearly differentiate the left half of a fundamental corpuscle from the right half of a fundamental corpuscle. Given that we can do this, surely God, who is omnipotent, can separate the left half of the corpuscle from the right half. If we can do it in thought, God can do it in reality. My view is that this simple and seemingly attractive argument rests on a confusion. Before offering a complete diagnosis of where it goes wrong, we can start just by showing that formal divisibility need not entail metaphysical divisibility.

First, Newton is very clear that the existence of conceptual parts does not entail the existence of metaphysical parts. Consider the following passage from *Tempus et Locus*: "But neither does Place argue the divisibility of a thing or a multitude of its parts, and on that account imperfection, since space itself has no parts."[66] Space manifestly has conceptual parts, places just are these conceptual parts of space. So the passage must mean that *despite* having conceptual parts space has

[64] *Principia* 796.
[65] *Principia* 795.
[66] McGuire 1978b, page 117.

no metaphysical parts. I think this is also the right way to read a passage from the General Scholium discussed in Chapter 5:

> Every sentient soul, at different times and in different organs of senses and motions, is the same indivisible person. There are parts that are successive in duration and coexistent in space, but neither of these exist in the person of man or in his thinking principle, and much less in the thinking substance of God. Every man, insofar as he is a thing that has senses, is one and the same man throughout his lifetime in each and every organ of his senses. God is one and the same God always and everywhere.[67]

There are temporal and spatial parts to souls. But these in no way threaten the unity or indivisibility of sentient souls. Mere conceptual parts do not entail metaphysical divisibility.

Of course, one might accept that in the instances of space, God, or souls, conceptual divisibility does not entail metaphysical divisibility, but allege that the case of bodies is different. For God and space there is a principled argument on offer. Even if we can conceptually differentiate parts of space from one another, there is no way to *separate* the parts of space from one another. Parts of space are immobile. Something similar will be true for God. Even if I can distinguish in my mind the part of God coextensive with a book from the part coextensive with the shelf that the book is on, these parts will not be separable. God is immutable so there can be no separating of God's parts or rearranging of them. Could a parallel argument be constructed for souls? What is there to stop God from cleaving the left half of a spatially extended soul from the right half? Some might say "nothing." But, for me at least, it is quite hard to envision what the outcome of such a procedure would be. And it is also hard to square that answer with Newton's claims about the indivisibility of persons in the block quotation just given.

Anyone made uncomfortable by the thought that God could unproblematically metaphysically separate one formal part of a soul from another is well-positioned to appreciate my view that the formal parts of bodies do not map onto metaphysical parts. The seemingly plausible argument that God could divide a fundamental corpuscle is compelling only if we hold a non-Newtonian view of bodies. On these conceptions of bodies, they are just made out of some "stuff," call it matter. And if a body is a mere "chunk" of matter then it can be hard to see why it would not be capable of infinite division, at least at God's hands. Reflecting on the fact that "matter" is a mass-noun can help make clear why this would be the case.

But this is *not* the Newtonian conception of bodies.[68] Newtonian bodies are powers situated in space. More specifically, they are powers located within

[67] *Principia* 941.

[68] My thinking here is partially indebted to the interpretation of divisibility in the "force-shell" matter theory of Boscovich and Kant developed in Holden 2004.

determined quantities of space. And one of the powers, mobility, is characterized as the power to move through space in a way that keeps the quantity and shape of the being stable. This is what ultimately explains why not even God could separate the right half of a fundamental body from the left half. Fundamental Newtonian bodies are not lumps or masses of matter that can be cut in half like a stick of butter. The fact that these fundamental bodies are defined in terms of their spatial distribution means that if they cease to maintain that spatial distribution they cease to exist. The powers are powers with a determinate size; they take up a determinate quantity of extension. That is what they are. Thus, there is no way to alter that size—say by division—while the power, understood as having a determinate quantity, continues to exist. Put differently, the only way for God to separate the left half of a corpuscle from the right half would be to destroy the powers that constitute the corpuscle and create two new sets of powers that are shaped differently and occupy smaller quantities of extension. Put differently again, given the way they are constructed, there cannot be metaphysical parts in these fundamental corpuscles.[69]

Note as well that physical indivisibility follows a fortiori from metaphysical indivisibility. So anyone unimpressed by the earlier considerations designed to show Newton believed fundamental bodies were physically indivisible but who sees the appeal of these arguments can feel free to substitute them in for the earlier ones. And, of course, those persuaded by both can aggregate them.

The discussion so far has focused on bodies. What about Gorham's claim that that Newtonian minds are utterly partless? Here I think we ought to insist that, whatever we say about physical or metaphysical parts, minds certainly must have formal parts. Newton is clear that minds are somewhere. He denies that a point is a real thing. So he must think that minds are extended. And having formal parts is a necessary consequence of being extended. If this is right, then minds and bodies are divisible and indivisible in just the same ways. They are both physically and metaphysically, albeit not formally, indivisible.[70]

Thus far we have seen theoretical and systemic reasons for rejecting Gorham's claim that Newton adopted a form of Cartesian dualism on which minds and

[69] Put in this way, the point is potentially misleading. The more precise claim would be that Newtonian bodies are not metaphysically *divisible* in the way that Cartesian ones are. The reason for clarification relates to the fact that, in some sense, the individual powers that constitute a Newtonian body are parts of it. What kind of parts? While united together they will not be *formally* separable from one another; they always overlap. But there is an interesting question of whether, having been created together, they are at a later time metaphysically separable by God. Could God, say, remove the power of penetrability from a body while allowing its other powers (mobility, mind-interactivity, etc.) to continue on as normal? Those who deny this to be possible would hold that they are mere intellectual, rather than metaphysical, parts. In any event, this is not the kind of divisibility at issue in Gorham's interpretation (or that of any one else), so I set it aside.

[70] Here it seems to me that Newton's ability to offer a reason for the metaphysical indivisibility of minds (namely, that God could not separate the formal spatial parts of the powers comprising them without destroying those very powers) represents a clear advance on the views of Henry More who had only offered the naked assertion that spirits, while spatially extended, were necessarily indiscerpible. For More's views on this point, see Reid 2012, page 190.

bodies are differentiated in terms of their divisibility. Before moving on, however, there is a key text Gorham uses in his interpretation which deserves some attention. The context is Newton's claim that God is infinitely extended in space: "Moreover, lest anyone should for this reason imagine God to be like a body, extended and made of divisible parts, it should be known that spaces themselves are not actually divisible."[71] Gorham's suggestion is that this provides evidence Newton is committed to the physical or metaphysical divisibility of bodies. I disagree that the passage can do this work.

First, it is not clear whether Newton is talking about a macro-body (e.g. a loaf of bread) or about a fundamental particle. For the passage to support Gorham's interpretation, we would need some decisive evidence that Newton was referring to the latter, rather than the former. One could easily hold that aggregates of fundamental corpuscles have parts, but that the corpuscles themselves lack physical or metaphysical parts. Second, and more importantly, it is not clear that in this passage Newton commits *himself* to the view that bodies are made of physically or metaphysically divisible parts. What Newton is doing in this passage is responding to a distinctively *Cartesian* worry. The Cartesian is worried that making God and minds spatial will make them divisible in the way that the Cartesian *herself* believes bodies are. Here, it is also worth remembering precisely where in the argumentative progression of *De gravitatione* the target passage occurs. Newton is still seeking to undo the errors of Cartesian conceptions of space. A central error of which is the conflation of space and body. And, at this stage, he has not yet offered his own positive theory of bodies. So, when Newton says that bodies are extended and made of divisible parts in this part of the text, we have reason to doubt he is speaking about *Newtonian* bodies or, more specifically, about Newtonian fundamental corpuscles. He could just be responding to the fact that Cartesians themselves believe bodies are infinitely divisible and that this a necessary consequence of their being extended.

What Newton then goes on to do is offer a way of conceiving how things can be diffused through space without having physical or metaphysical parts. And he does this by reference to the way in which moments of time lack physical or metaphysical parts but are, nonetheless, in space. Admittedly, in the course of this argument, Newton does contrast the way moments of duration relate to space with the way in which bodies relate to space. He writes that "the relation of duration to space is very different from that of body to space."[72] But the contrast he wants is *not* that duration is entirely partlessly extended in space whereas bodies have (some kind of) parts. Instead, the next sentence specifies that "we do not ascribe various durations to the different parts of space, but say that all endure simultaneously." The point is just that whereas moments of duration occupy *all* parts of space, different bodies

[71] *DG* 40.
[72] *DG* 40.

occupy different parts of space. This explains why Newton begins the next paragraph with a sentence meant to draw out this contrast: "The positions, distance, and local motions of bodies are to be referred to the parts of space."[73] And this is all Newton needs in the dialectical context, he just needs that the parts of duration are not physically divisible, not that duration lacks parts. Further evidence for this line of thought stems from a fact about the manuscript pointed out by Janiak.[74] Specifically, where Newton wrote that "spaces themselves are not *actually* divisible"—in the Latin: "spatia non esse actu divisibilia"—the word "actu" or "actually" was an interlinear insertion added after the sentence had first been written without it. This indicates that Newton saw the need to be precise in indicating that it was only physical or metaphysical, and not conceptual, divisibility that was the focus of the passage.

Given all of this, I do not see *divisibility* as marking a fundamental difference between minds and bodies. Another candidate for a potential differentiator is *extension*. Could it be that bodies are extended in space differently than minds? And could this be what characterizes Newton's dualism?

Again, some preliminary distinctions will be useful. A substance can be extended by having *partes extra partes*. For the moment, we can set aside the question of just what kind of parts are outside of parts. Alternatively, a substance can (at least according to some) be extended *holenmerically*, such that the whole is in the whole and the whole is also in each part.[75] The substantial form of a human, in standard Scholastic accounts, was understood to be extended in this way throughout the body. And God is another prime candidate for a substance that is holenmerically extended.

Andrew Janiak has claimed that Newtonian minds and bodies differ in that while both are extended, bodies are extended such that they have parts whereas minds are extended but partless.[76] I take him to mean that bodies are extended *partes extra partes* whereas minds are extended holenmerically. This is supported by Janiak's claim that Newton, rejecting Cartesian dualism, was following Henry More in thinking there are two different ways of being extended. Hylarie Kochiras, by contrast, argues that both bodies and minds are extended holenmerically.[77] And Ed Slowik argues that all Newtonian substances are extended *partes extra partes*.[78]

Ultimately, I think questions surrounding Newton's attitude toward holenmeric extension are tremendously difficult.[79] Simply put, there is precious little text and

[73] *DG* 40.

[74] See Janiak 2000, footnote 76.

[75] For helpful background on this distinction and several related ones as they feature in Newton's thought and context, see Kochiras 2012; Reid 2012, pages 227–229; McGuire and Slowik 2012.

[76] Janiak 2010, page 667. See also Slowik 2013, pages 439–440.

[77] See Kochiras 2013.

[78] Slowik 2016, ch. 2, sec. 4.2.

[79] I therefore sympathize with McGuire and Slowik, who in the course of a helpful treatment of these issues claim that "his position on holenmerism is . . . difficult to pin down, and in fact constitutes the most difficult aspect of his spatial metaphysics." McGuire and Slowik 2012, page 292.

the relevant texts we do have are difficult to interpret decisively. In that sense, I am hesitant about all of the interpretations on offer. That said, I think Kochiras and Slowik are correct to insist on a kind of parity. This is for two reasons. First, whether they are extended *partes extra partes* or holenmerically, bodies and minds are best understood as extended in the same way. The reason is that, while they might be composed of different powers, minds and bodies are fundamentally the same type of thing: collections of powers located in space. Second, in keeping with the discussion from above, it does not seem that Newton saw either minds or bodies as having either physical or metaphysical parts. And anything extended is going to have (mere) formal parts. So, again, it seems that there is a kind of parity. This means that we should not, as Janiak does, characterize Newtonian dualism as a thesis about the different ways in which different types of substance relate to space.

We have seen above that Newton's dualism is not Cartesian. Janiak had suggested it might be Morean. And I have just argued that this is also incorrect. In connection with this, we can also note that More's metaphysical views underwent a number of considerable shifts throughout the course of his career.[80] And while it is true, as Janiak highlights, that at one stage in his career More held spirits to be holenmerically extended, it is equally important to note that he ultimately abandoned this view. And he abandoned the view precisely because he came to see holenmeric extension as unintelligible. Newton was familiar with texts, like More's *Enchiridion metaphysicum*, which showcased this change of heart. But while we can establish that Newton likely encountered these Morean anti-holenmerism polemics, it is difficult to say whether or not he was persuaded by them. On my view, there is something to be said for the points made by both Kochiras and Slowik in support of their opposing positions. Either way, I think we have reason to understand Newton as thinking that bodies and minds are extended in the same way, whatever way that is.

Quite apart from extension, there is another way of understanding Morean dualism. More very often worked to offer a fundamental contrast between minds and bodies along the lines of *active* and *passive* substances.[81] *CPQ* does indicate that Newton had been reading More very closely around the time it was composed and this is relatively unsurprising given his Cambridge context. So it is natural to ask whether he adopted *this* Morean view. Newton obviously believed minds were active and capable of spontaneous thought and volition. And a number of commentators—in the course of addressing issues unrelated to Newton's dualism—have claimed that Newton was firmly wedded to the passivity of bodies.[82] So the view deserves consideration.

[80] For detailed discussion see Reid 2012.

[81] This having been said, for a careful discussion of More's dualism that situates activity and passivity downstream from penetrability and discerpibility, see Reid 2012, pages 187–188.

[82] E.g., Heimann and McGuire 1971, pages 235–236; McMullin 1978, pages 29 and 53; McMullin 1985, pages 51–52; Rogers 1982, page 225; Wilde 1982, page 101; Guerlac 1983, page 224; Leshem 2003, page 68; Ducheyne 2009a, page 238; Ducheyne 2012, page 37.

Ultimately, I am disinclined to see Newton's dualism as Morean in this sense. The chief reason is that Newton does not seem particularly attached to the passivity of bodies. At a minimum, he never emphasizes the passivity of bodies in the way that More does.[83] But there are also two positive reasons for thinking Newton sees bodies as active. The first is that Newton identifies bodies as substances and says that part of what it is to be a substance is to act.[84] The second relates to the reconstruction of bodies in *De gravitatione* that I have offered. On this story they just are *powers*, so will be active to some degree. Of course, one might work to understand the "actions" of bodies-qua-substances as a function of their inertial forces. This might bring them more in line with passive Morean bodies. And one could also opt for an alternative understanding of the powers of impenetrability and mobility in a way that makes bodies passive. But, I think, it will be very difficult to do the same with the power to arouse perceptions in human minds. This seems like a distinctly active power.[85]

There is one other possible sense in which Newton's dualism might be Cartesian or Morean. And it is one so obvious that Newton himself must have been intensely aware of it. For Descartes and for More, the human soul is naturally immortal whereas bodies are corruptible. As will be discussed in the next chapter, this is a position that Newton rejects. So immortality will also fail to serve as a key differentiator for Newtonian minds and bodies.

7.7. A Positive Proposal

In the previous sections, we looked at a number of ways in which Newton's dualism might be characterized. It is not Cartesian because Newton allows that minds are spatially extended, because (fundamental) bodies are not physically or metaphysically divisible, and because the soul is not naturally immortal. Newton's dualism also differs from the early views of Henry More, on which mind and body are extended in different ways. And I have suggested that it is not entirely clear whether or not he agreed with the later More, who understood both as being extended *partes extra partes*. At any rate, Newton's dualism is not Morean because he

[83] There is one passage in a draft of the Queries where Newton does write that "Matter is a passive principle & cannot move itself." CUL Ms. Add 3970.3, 619r. But the parallel content in these drafts makes the same point by saying that it is the *vis inertiae* that is the passive principle. See 254r, 281r, 620v. I propose that this latter claim is closer to Newton's true view. Dhondt and Ducheyne have argued that the 254r passage focused on the *vis inertiae*, in particular, postdates the version of the claim focused on matter. See Dhondt and Ducheyne 2021, sec. 2 for discussion. An alternative strategy for disarming this passage is to look at the larger context of the claim and argue that Newton is not speaking in his own voice, but is instead singling out a Cartesian principle for criticism. This line of thought is pursued in Schliesser 2021, pages 55–58.

[84] *DG* 36.

[85] A similar analysis is offered in Stein 2002, pages 266–267.

rejects the absolute passivity of matter and—again—because he denies the natural immortality of the soul. So how should we characterize Newtonian dualism?

To my mind, the question is an important one because it helps to bring out some of the key features of Newton's metaphysics. I suspect that we often think of dualism in an *oppositional* way. It is meant to be a theory on which there are two fundamentally distinct kinds of things, two kinds of things with mutually exclusive essences. This is part of what powers Descartes' argument for the real distinction. And the point is very explicit in the preface to the *Meditations* where he writes that "the nature of mind and body are not only different, but in some way opposite."[86] Similarly, More's *Divine Dialogues* appeal to the "rule of Contraries" in characterizing matter and spirit.[87] What is intriguing, and perhaps appealing, about Newton's basic framework for thinking about secondary substances is that it will never offer a dualism of quite this type.

Newtonian secondary substance have no essences or natures. They just are collections of powers. And God had free choice in selecting which powers to create and unite together. Any set of mutually compatible powers can form a secondary substance.[88] As it happens, or at least as far as we can discern, God created two types of these things. Their "essential natures" are exhausted by the powers that comprise them. And this ensures that they are distinct. But it also means that characterizing the difference between them in some way other than by enumerating the powers and noting that they are not shared between the two types of substances is misguided. Because we happen to find impenetrability and law-mobility united together we might normally categorize these as "bodily" powers.[89] And because we happen to find sensation and volition united together we might normally categorize these as "mental" powers. But there are no necessary connections to be had between these powers. And they might have been assembled together into different clusters.

So my positive proposal has it that Newton's dualism can be characterized. But the radical nature of his metaphysics means that there is no key defining feature shared by all powers of bodies and no key defining feature shared by all powers of minds. And, consequently, there is no deep opposition between the "corporeal" and the "mental."

One might think that this would give Newton a competitive advantage in the quest to solve the well-known issues relating to mind-body interaction. In one sense, it does. And Newton does solve the mind-body problem. Of course, in some other important sense, Newton's solution may appear to be underwhelming. God simply endows bodies with the power to affect minds and be affected by them. And

[86] *CSM*, II, 9–10. Underlining added.
[87] More 1668, page 124.
[88] For more on this point, see Connolly 2024, sec. 5.
[89] See Newton's comments on homogeneity in the Definitions at *NCS* #116. See also McMullin 1978, page 9, and the helpful discussion of Maupertuis' approach to these issues in Downing 2012.

God simply endows minds with the power to affect bodies and be affected by them. The problem is solved by brute force, or at least by divine fiat. Nonetheless, I think there is still value in holding on to the sense in which Newton did succeed. This is because his success is perceptive. Newton realized that the concepts of mind and body used by Descartes and others rendered the problem insoluble. Innovative and non-oppositional concepts of minds and bodies were needed to explain how mind-body interaction might even be possible.

7.8. The Third Law of Motion

Newton was committed to the view that minds and bodies do, in fact, interact and was highly dismissive of positions that problematized this. Consider, for example, a claim made in a letter to Conti: "His [Leibniz's] Harmonia præstabilita is miraculous & contradicts the daily experience of all mankind, every man finding in himse[l]f a power of seeing with his eyes & moving his body by his will."[90] Cartesian dualism, notoriously, faces difficulties in clearly explaining how it is that minds and bodies might interact. Often, these difficulties are discussed under the rubric of the so-called heterogeneity problem: Given that minds and bodies are so radically different, how is it that they can interact with one another?

We have just seen Newton tackling one aspect of this heterogeneity problem. Cartesian minds and bodies have different ways of relating to space. For Newton, this Cartesian position was unacceptable in part because it held that the mind was not spatially extended and this "renders its union with body thoroughly unintelligible and impossible."[91] So Newton clearly diagnosed Descartes as being subject to this heterogeneity problem. One key step in resolving these difficulties involved allowing both minds and bodies to be extended and allowing that minds and bodies might be co-located. Importantly, however, it seems that Newton realized that this might be a necessary, rather than sufficient, step in solving the problem.

A passage from *CPQ*, while not directly discussing mind-body interaction, illustrates Newton's sensitivity to other aspects of the heterogeneity problem. In this passage, Newton is arguing that the motion of a body must be due to a continued impressed force. That force must either be "some corporeal efflux or incorporeal one."[92] If it is an incorporeal efflux, it must either be a spirit or quality. Important for our purposes is Newton's argument that it cannot be an incorporeal spirit: "If [the motion is continued] by a spirit, how comes the spirit to be so easily united to the body, and not to slip through it."[93] This suggests Newton recognized that

[90] *NCS* #1187.
[91] *DG* 46.
[92] *CPQ* 407.
[93] *CPQ* 407.

spatial extension and proximity were not enough for mind–body interaction. And this gives some insight into why Newton, ultimately, "solves" the problem of mind–body interaction through brute force. He just makes it one of the constitutive powers of bodies that they can affect and be affected by minds. As for the mind, consider part of the quotation from "Of God" referenced above: "I can conceive all my owne powers (knowledge, activating matter etc) without assigning them any limits."[94] It is revealing that the power to activate matter just *is* one of the mind's powers here. Just like the power to receive perceptions from bodies, the volitional power to control the body is simply a constitutive power of the mind. In this sense, Newton managed to side-step one serious problem facing dualists who embrace interaction.

But the heterogeneity problem is not the only one facing Cartesian interactionism. There are also well-known problems relating to the conservation of motion.[95] According to Descartes, there is a conservation of quantity of motion in the universe. This makes it very hard to understand how it is that our minds could cause new motions to begin in our bodies. Suppose I am at rest on the sofa but then, desiring to get a beer from the fridge, I form the volition to stand and go into the kitchen. Upon executing the volition and rising from the couch it seems as though some new motion has been added to the universe at the behest of my will. This violates the conservation of quantity of motion.

Importantly, Newton will not face this problem. Indeed, there is passage in a draft of the Queries where he seems to clearly affirm that motions added to the body by the will proceed without the diminution of motion elsewhere in the universe.[96] But, as Alan Gabbey has outlined, it seems Newton may face a problem closely related to the Cartesian one.[97] Unlike Descartes, Newton is not committed to the *global* conservation of quantity of motion in the universe. But the third law of motion in the *Principia* does commit him to actions having equal and opposite reactions. This raises a puzzle about what happens when, as in the scenario above, my will causes me to rise from the sofa. This seems to be an action of the mind on the body; but where is the equal and opposite reaction? What would such a reaction even be? Gabbey's verdict is as follows: "There is no evidence that he [Newton] was aware of the mismatch between his third law and his inviolable belief in the power of the human mind to intervene in the mechanism of the world. This parallel between Descartes and Newton points to the incompatibility between

[94] Dibner Ms. 1031 B, 4v.

[95] For a recent treatment, including a discussion of various solutions open to Descartes, see Archer and Ragland 2022. In connection with Newton, it is intriguing to note that Henry More had raised this as a difficulty for Descartes in a letter of 23 July 1649.

[96] CUL Ms. Add 3970.3, 619r.

[97] See Gabbey 2016. As Gabbey notes, Boscovich, MacLaurin, and Hume all seem sensitive to the issue. See Gabbey 2016, pages 443–444.

the doctrine of human freedom of action and the doctrine of the inviolate rule of physical law."[98]

While I think it plausible that Newton never considered the possibility that there might be a problem here, I am less convinced than Gabbey that there is a genuine problem for Newton on this point. Careful attention to the wording of Newton's three laws reveals an asymmetry between the first two and the third. The first law says that bodies maintain their state until compelled to change by an impressed force. The second law says that changes in motion are proportional to the force impressed. But these two laws are silent with regard to the origin or nature of the impressed force. By contrast, the third law begins by claiming that for any action there is an opposite and equal reaction, but it then goes on to offer an alternative formulation of the claim. The two formulations are separated by a "sive" so are presumably intended to be equivalent. The second claims that "the actions of two bodies upon each other are always equal and always opposite in direction."[99] What this suggests is that the scope of the third law is intended to be limited.[100] It is only intended to apply to body–body interactions.[101] So no violation of the third law occurs if a soul exerts a force on a body without that body exerting a force on it in return. Further confirmation of this idea comes from a draft of the Queries in which Newton writes the following: "Life and will are active Principles by which we move bodies, & thence arise other laws of motion unknown to us."[102]

One reason this might be significant is that Newton is often at pains to stress the similarity between God's action in the world and the action of our wills on our bodies. God certainly can act in the world to, for example, move a stone. We would never want to say that in so doing the stone exerted a force on God. And, insofar as our agency is modeled on or similar to God's, the same will be true for us. Gabbey is right that the doctrine of human freedom of action was hugely important to Newton. As Newton claims, "{A sinn}er does evil not because he cannot do all things that he will but becaus he will not do all things that he can . . . His wil against evil is determined not by any blind impuls of his nature but by free election."[103] Or, even more explicitly, "A man may bring plausible arguments against

[98] Gabbey 2016, page 445.

[99] *Principia* 417. Underlining added.

[100] Iliffe, though not considering the specific problem of reconciling free will with conservation laws, offers a version of this view: "The fact of free will and the capacity for self-motion were always held by Newton as evidence that the normal laws of motion had application over a limited domain." See Iliffe 1995, page 450.

[101] It is perhaps notable that Descartes' third law of motion, unlike his first two, is also similarly restricted to interactions between bodies. See *CSM*, I, 242. In connection with this it can be noted that, while the problems facing Descartes and Newton are not identical, my response to the problem for Newton bears a certain kinship to the response Garber outlines on behalf of Descartes. See Garber 1983.

[102] CUL Ms. Add 3790.3, 619r. Underlining added. Similar wording appears at 620v and, canceled, at 252v and 254r.

[103] Keynes Ms. 2, Av. The observation at Mandelbrote 2006b, page 339, that Newton had an antipathy toward predestinationism is interesting in connection with this remark.

the power of the will but I'le believe experience."[104] But Gabbey is wrong to suggest that this commitment comes at the cost of abandoning the inviolate rule of physical law. Physical laws are precisely that: physical. And Newton's wording indicates that mind–body interaction simply falls outside the scope of the third law.

7.9. Conclusion

This chapter began with a look at Newton's claims about minds in *CPQ* and *De gravitatione*. I argued that, rather than endorse monism or agnosticism, both of these texts show Newton endorsing a dualist position. *CPQ* is also revealing insofar as it shows Newton working to differentiate and understand our various mental faculties. I argued that, given his core commitments regarding the ontology of created substances, Newton must understand these mental faculties as constituting our minds, akin to the way in which bodies are also constituted by collection of powers. This, intriguingly, means that Newton's dualism is different from other, Cartesian and Morean, dualisms on offer in the period. Finally, I considered what Newton's commitment to the third law of motion might mean for his views on mind–body interaction. The next chapter sticks with the theme of Newtonian minds. But it changes its focus to the later stages of Newton's career and asks whether or not his views underwent substantial change or development.

[104] CUL Ms. Add 3970.3, 619r.

8
Mind–Body Agnosticism and Sensoria

8.1. Introduction

The previous chapter argued that, at least in the early and intermediate stages of his career, Newton embraced a form of dualism. This chapter will explore the possibility that Newton, later in his life, may have come to adopt a kind of agnosticism with regard to the question of whether the powers constituting the body and the powers constituting the mind form one substance or two. I will argue that this agnosticism was made possible by key elements of his metaphysical worldview and may have been salient to him in virtue of his close association with John Locke. Exploring the contours of such a position will provide some useful insight into the unique nature of Newton's views.

Ultimately, while a form of agnosticism was open to Newton, I do not believe he ever adopted the view. In part, this is because there is simply not enough textual evidence to support the claim that he did. But it is also because of Newton's commitment to the claim that space is (as it were) God's sensorium. I will argue that, properly understood, these claims show Newton's lasting commitment to dualist models of mind–body interaction. An examination of Newton's claims about the divine sensorium also provides a fitting final topic for the book. This is because they showcase Newton's metaphysical views about God, bodies, minds, and the way they all relate to one another.

8.2. The Question of Agnosticism

Recall from the previous chapter that Newton's dualism was strikingly *non-oppositional.* Rather than see minds and bodies as fundamentally different or heterogeneous kinds of things, Newton actually understood them as being similar in nature: Both were understood as collections of powers located in space. Further, there was nothing about the powers themselves that dictated which ones must be united together beyond the free choice of God. On a standard Scholastic view, for example, God would create a substance by actualizing an essence whose nature was already (in some way) determinate. It was by actualizing the essence of, say, a cow that the powers of mooing, walking, digesting grass, and so forth were created. This was not Newton's picture. Newton's God was able to create substances by choosing powers, as it were, *à la carte.*

Newton's Metaphysics of Substance. Patrick J. Connolly, Oxford University Press. © Patrick J. Connolly 2025.
DOI: 10.1093/9780191959240.003.0008

In his commitment to dualism, as I have reconstructed it, Newton believed God had done this in two ways. One set of powers God had created and bound together included things like impenetrability and mobility. Another set of powers God had created and bound together included things like memory and volition. This is, of course, one possibility. But the structure of Newton's views gives rise to a further possibility; rather than being formed by the interaction between two separate substances, humans might instead be the aggregate of "bodily" and "mental" powers bound to one another in a single substance. It is unclear just how one might go about finding out which of these possibilities is actual. Hence, Newton had a clear route to a principled form of agnosticism regarding human ontology.

There are some pieces of evidence that suggest Newton himself might have been aware of, and even favorably inclined toward, just such a view. Consider, for instance, the rather cautious language used in some passages from Query 28, a relatively late text:

> Is not the Sensory of Animals that place to which the sensitive Substance is present, and into which the sensible Species of Things are carried . . . that there they may be perceived by their immediate presence to that Substance?[1]

> . . . Images . . . carried through the Organs of Sense into our little Sensoriums, are there seen and beheld by that which in us perceives and thinks.[2]

Newton here seems to carefully eschew language of minds, spirits, or souls. He opts for the rather guarded language of sensitive substances and that which perceives and thinks. Further, this phrasing was carefully weighed, the same language shows up many times in the drafts for this text.[3] To this, we can add a report from Conti:

> Mr Newton only speaks of the soul and the body in relation to phenomena; he protests that he is perfectly ignorant of the nature of these two beings. By body he only means that which is extended, impenetrable, heavy, etc., by soul that which thinks, that which feels in us etc. He says he doesn't pursue it further.[4]

Finally, there is something relevant in one of David Gregory's memoranda from May of 1694. He reports that, according to Newton, "Not a separate existence of the soul, but a resurrection with a continuation of memory is the requirement of religion."[5] This, of course, is compatible with a certain level of nescience regarding the ontological makeup of persons.

[1] *Opticks* 344.
[2] *Opticks* 345.
[3] See CUL Ms. Add 3970.3, 242r, 243r, 247r, 252r, 271r, 286r, and 619r.
[4] Leibniz, 1875–1890, volume 3, page 653.
[5] *NCS* #446. Admittedly, evidence from Gregory's memoranda should be regarded as defeasible. There are instances in which Gregory attributes views to Newton that are in tension with Newton's

Given these reports on Newton's views and some suggestive bits of text, we have good reason to consider the possibility that Newton came to consider or adopt a form of mind–body agnosticism.

8.3. Newton and Locke: Mind–Body Agnosticism

The Gregory memorandum just quoted is particularly notable for its deep Lockean resonances.[6] Locke's views on personal identity, its relationship to moral responsibility and ultimate divine judgment, and the ontology of human beings are notoriously difficult and controversial. But Locke arguably (a) held an agnostic position on the ontology of human beings, (b) denied that a "separate existence of the soul" or, indeed, the continuous existence of any substance was required for the ends of morality and religion, and (c) crafted his views on personal identity with a clear eye toward theological considerations.[7] Whether Locke's claim that continuity of consciousness is the hallmark of personal identity is best cashed out in terms of memory is now controversial, but this was certainly a prominent interpretation in the early reception of Locke's work. Given this, we have some reason to consider whether Locke and Newton discussed issues relating to human ontology and whether Locke might have influenced Newton.[8]

Locke and Newton met some time around 1690, after Locke's review of the *Principia* and near the time of the *Essay*'s first publication. They became close relatively quickly and maintained a friendship up through Locke's death in 1704. Newton twice visited Locke at Oates, Locke visited Newton once at Cambridge, they had ample opportunities to meet while both in London, and they exchanged a number of letters.[9] We know the two discussed themes relating to the natural philosophy of the *Principia*.[10] But we also know they discussed issues that might bear on minds, bodies, and their relationship. Two of these stand out. First, we have a response from Newton to a now lost letter from Locke in which he reports on the afterimage experiments he conducted as an undergraduate.[11] This was apparently

written works or other commitments. In this instance, however, I can see no prima facie reason for doubting the veracity of the report. See Iliffe 2017, page 307, for the suggestion that this claim of Gregory's aligns with many of Newton's other views.

[6] Dempsey 2006, page 428, also comments on this.

[7] For (a) see Jolley 2015. For (b) and (c) see Boeker 2021.

[8] There is a common tendency to focus on questions of whether and how Newton influenced Locke. In asking about influence flowing in the opposite direction, I am partly inspired by Michael Jacovides' arguments that Newton's Rules for the Study of Natural Philosophy may be indebted to Locke's claims about primary and secondary qualities. See Jacovides 2017, pages 90–91. On this point, see also McGuire 1968b and, more generally, Rogers 1979 and Ducheyne 2009b.

[9] Westfall 1980, pages 488–493 and 590–591.

[10] See, for instance, Brackenridge 1993.

[11] *NCS* #365.

prompted by a question Locke posed concerning a passage in Boyle's *Experiments and Considerations Touching Colours*.[12] Newton concludes with the following:

> This story I tell you, to let you understand that in ye observation related by Mr Boyle, the man's phansy probably concurred wth ye impression made by ye sun's light to produce that phantasm of ye sun which he constantly saw in bright objects. And so your question about ye cause of this phantasm involves another about ye power of phansy wch I must confess is too hard a knot for me to untie.[13]

This allows the probable conjecture that Locke had asked about a passage in the book where Boyle discussed his experiences of looking at the sun using a shaded glass and a telescope and reported on the changes to his color phenomenology after looking away. Boyle also reported on the experiences of others in similar situations.[14] What is intriguing is that Newton is here arguing that the phenomenon in question should be explained not just with reference to the body's impact on the mind (not just in terms of the motions in the ocular nerve and brain producing a phenomenological experience of different colors), but also with reference to the mind on the body (with reference to how we direct our imagination and the consequences it can have on the physiology producing our sensations). Put differently, the discussion turns from mere perception to mind–body interaction. Second, much of Locke and Newton's mutual conversation was dominated by theology and Scriptural issues. And it is, of course, likely that they discussed issues relating what beliefs about human ontology were "the requirement of religion" in this context.

Quite apart from their personal interactions, Newton might have been influenced through his reading of Locke's work. Newton's library contains a number of works by Locke.[15] Most important is a copy of the first edition of Locke's *Essay* bearing two corrections to the text in Locke's hand and significant signs of dog-earing.[16] There is also a passage from a draft preface to the second edition of the *Principia* that seems, to my mind, to show Newton comfortably operating with a broadly Lockean epistemology:

> What is taught in metaphysics, if it is derived from divine revelation, is religion; if it is derived from phenomena through the five external senses, it pertains to physics; if it is derived from knowledge of the internal actions of our mind

[12] Newton does not seem to have owned this work, but Locke did. See Harrison and Laslett 1971, page 93. Nonetheless, Newton was almost certainly familiar with the work and his afterimage experiments were likely inspired by it. See Tamny 1976, page 161 and Iliffe 2017, page 112.
[13] *NCS* #365.
[14] Boyle 1664, pages 15–19.
[15] *NL* #966–976.
[16] *NL* #967.

through the sense of reflection, it is only philosophy about the human mind and its ideas as internal phenomena likewise pertain to physics.[17]

One final link between Newton and the *Essay* comes from Conti who reported in a letter that "there is no doubt that everything that is most solid and most original in Locke's *Essay concerning Human Understanding* comes from him [Newton]."[18] This claim, as it stands, is obviously false. There is extensive draft material for the *Essay* that predates Locke and Newton's meeting. What matters for current purposes is that the claim shows Conti, who knew Newton fairly well, thinking it was plausible to associate Newton closely with the main lines of thought in Locke's *Essay*.

Given all of this, it is worth sketching out the deep resonances between Locke's actual and Newton's possible agnosticism.[19] One place to start is with what I characterized, in the previous chapter, as Newton's "non-oppositional" approach to minds and bodies. For Newton, unlike for some other dualists, the mere fact that some substance was a body would not necessarily preclude it from also being a mind. Locke has precisely the same attitude. In Book IV, ch. 3, sec. 6, Locke allows that we have ideas of bodies and ideas of minds but that there is simply nothing in those two ideas that would allow us to perceive a contradiction in the idea of a thinking material object. And, as with Newton, this is funded by a fairly "lightweight" approach to essences. Figures like Descartes believed we could know the essences of minds and bodies and therefore determine that they had incompatible attributes. Newton, of course, rejected the idea that secondary substances might have these sorts of essences. Locke denied that we had any knowledge of the *real* essences of substances. This meant that, in Newtonian fashion, our knowledge of them and our sorting of them into kinds was dependent on their observable features, on what Locke would call their *nominal* essences. And this makes it at least epistemically possible that an object could have the nominal essence of a body while still thinking. Consider what Locke tells Stillingfleet: "The idea of matter is an extended solid substance; wherever there is such a substance, there is matter, and the essence of matter, whatever other qualities, not contained in that essence it shall please God to superadd to it."[20] Solidity and extension (on Locke's telling) are sufficient for something to be a body. And it will continue to be a body even if it also has other, mental, powers as well. So neither Locke nor Newton sees anything in the mere ideas of bodies and minds that would preclude a single substance from being both a body and a mind.

[17] *Principia* 54. Ducheyne 2012, pages 256–257, also notes the Lockean tenor of this passage. See also Ducheyne 2009b, pages 259–262.

[18] The letter is transcribed at Badaloni 1968, page 239.

[19] For more on Locke's agnosticism, thinking matter, and superaddition as well as citations to relevant secondary literature, see Connolly 2015b and 2019.

[20] Locke 1823, volume 4, page 460.

The passage just quoted from Locke's letter to Stillingfleet points toward a second deep affinity between Locke's agnosticism and Newton's thinking about these issues: divine omnipotence and powers. Locke consistently invokes God's omnipotence, especially in connection with the weakness of human understanding, as an argument for the possibility of a thinking material substance. More specifically, he often claims that there would be no obstacle to God "superadding" a power of thought to a material object. This idea, that God can endow objects with powers at will and that God is not under any obligation to create sets of powers with any particular profile, was also one clearly endorsed by Newton. So again, there is reason to think that in getting to grips with Locke's views Newton might have seen the possibilities for a principled mind–body agnosticism.

8.4. Newton and Locke: The Root of Morality

The previous section argued that Newton might have been aware of the kind of agnosticism open to him and even favorably disposed toward it in virtue of his close association with Locke. And I indicated that thinking about Locke's views was helpful for thinking through the logical space open to Newton. One challenge to this idea comes from a recent paper by Liam Dempsey which argues that Locke and Newton had deep and fundamental differences on precisely these issues and, further, that these differences led directly to one of the lowest and most acrimonious moments in their friendship.[21] I want to offer an alternative interpretation of these events.

The setup is as follows. In September of 1693, Newton wrote a letter of apology to Locke.[22] We do not know what passed between them in the immediate run-up to this letter, but clearly it was not good. Newton, in part frustrated because he believed Locke had "endeavoured to embroil [him] with woemen" even confesses to having wished Locke dead. Central to our concerns is Newton's apology for "representing that you struck at ye root of morality in a principle you laid down in your book of Ideas & designed to pursue in another book & that I took you for a Hobbist."[23] It is very clear that by Locke's "book of Ideas" Newton is referring to the *Essay*. The puzzle is about which aspect of the *Essay* Newton found so offensive.

Dempsey's view is that Newton, to whom he attributes "an idiosyncratic brand of substance dualism," is reacting to what he perceives as Locke's materialism, a materialism dangerously close to that of Thomas Hobbes.[24] Dempsey further argues that Newton was right to be worried, in the sense that Locke's actual views

[21] Dempsey 2017.
[22] *NCS* #421.
[23] *NCS* #421.
[24] Dempsey 2017, page 454.

were, in fact, very close to being materialist. For our purposes, we can largely set the questions of Locke interpretation to one side.[25] Whatever the facts are about Locke's actual position, it is certainly true that a number of contemporary readers did see the *Essay* as endorsing some kind of materialism. So, it is at least conceivable that Newton could have read the *Essay* in this way. And if he did, this would endanger my suggestion that Newton might have been open to a kind of agnosticism on these issues and might have been favorably inclined toward the view at least in part due to his association with Locke.

I am skeptical that Dempsey's is the right understanding of the acrimonious episode between Locke and Newton. My skepticism has two main sources. First, given the context in which Newton made these accusations, it is not clear that we should take them to have any significance at all. Second, there is a better explanation of Newton's accusations available.

Why might we want to be hesitant about using the letter mentioned above to gain insight into Newton's views? The reason is that the events it alludes to occurred at a time of great mental and emotional instability for Newton. The historical record makes it difficult to determine precisely what happened, but the extant sources all indicate that Newton was seriously unwell.[26] Newton's belief that Locke was a Hobbist came at a time when Newton was suffering and having serious difficulties sleeping. By his own telling, he wrote the letter after having been awake for something like ninety-six hours.[27] Given this, and given Newton's subsequent downplaying of the letter's significance and request that it be forgotten, the wisest course might be to simply ignore the letter as a valuable source of information. Dempsey does have some sympathy for this line of thought and acknowledges that Newton's frame of mind presents problems for understanding what he found so offensive in Locke's thought.[28] But he nonetheless maintains the most likely explanation is that Newton was responding to what he saw as Locke's materialism.

If we are intent on reconstructing Newton's understanding of Locke on the basis of this letter, I think we would do better to see Locke's views on morality, rather than his views on human ontology, as the target of Newton's worry. The part of

[25] For some critical consideration of Dempsey's interpretation of Locke, see Rickless 2020, sec. 4. My own view, briefly articulated above, is that Locke had developed a principled agnosticism on the issue and that accusing him of crypto-materialism is a mistake.

[26] See Westfall 1980, pages 533–540 and Newman 2019, pages 393–394.

[27] See *NCS* #426. The fact that Newton believed Locke "endeavoured to embroil [him] with woemen" is interesting in connection with Newton's lack of sleep during this period. It may have been that Newton believed Locke was involved in some plans surrounding a potential marriage match for him. But some of Newton's claims about the ways in which early Christian monks became corrupted suggests an alternative possibility: "The want of sleep & due refreshment disorders the imagination and at length brings men to a sort of distraction & madnesse so as to make them ~~think they~~ have visions of weomen conversing with 'em ~~& sitting upon their knees~~ & think they really see & touch them & heare them talk." "Paradoxical Questions concerning the morals & actions of Athanasius & his followers," UCLA Clark Library Ms. 1961.009, 75r.

[28] Dempsey 2017, pages 455 and 456.

Newton's charge that concerns us is threefold. First, that Locke struck at the root of morality in a principle laid down in the *Essay*. Second, that Locke aimed to pursue that principle further in another book. And third, that Locke was a Hobbist. To my knowledge, there are three suggestions for what might be going on here. First, Newton may have been responding to Locke's attack on innate ideas.[29] Second, as Dempsey argues, he may have been accusing Locke of materialism.[30] Third, he may have been offended at aspects of Locke's metaethics and moral psychology.[31] I will argue that this third proposed explanation is the best one.

The charge of Hobbesianism was an expansive one in the early modern period and covered a multitude of sins.[32] Consider the wonderfully inclusive sense in which John Edwards accused Locke of being a Hobbist. He claims that Locke follows Hobbes:

> [I]n his Opinion of *the Necessity of only One Fundamental Article of Christian Faith*, and in his Notion of *Thinking matter*, and particularly in that of the likelihood or possibility of the *Materiality of Humane Souls*, and of their tendency (on that Principle) to *Mortality*, and in his *Contempt of some parts of the Holy Scripture*, and in his avow'd disbelief of the *Resurrection of the same Body*, and in his ridiculing of the *receiv'd Explication of the Doctrine of the Holy Trinity*, and in his general favouring of *Scepticism* and *Infidelity*, and in his denial of *Natural* and *Inbred Notions*; [and] in his belief of *the precarious and arbitrary nature of Mortality*.[33]

There is no shortage of charges here. By my count, Edwards sees Locke as guilty of Hobbism for eleven (or perhaps twelve) distinct reasons ranging from credal minimalism to anti-nativism. The point is just that a certain amount of caution is needed in moving from Newton's accusation of Hobbism to the conclusion that the accusation was motivated by Locke's (alleged) acceptance of any one particular doctrine.

I think the idea that Newton would have been deeply offended at Locke's anti-nativism can be safely discarded. It is true that Hobbes also denied innate ideas, but this was hardly a hallmark Hobbesian position. More importantly, Newton himself was strongly opposed to nativism. A draft of one of the rules for philosophizing for the third edition of the *Principia* contains the following: "But I do not perceive that

[29] This position has been taken by Newton's early biographer Brewster and by de Beer, the editor of Locke's correspondence. See Brewster 1860, volume 2, page 100 and Locke 1976–1989, volume 4, page 727. Collins 2020, page 329 also adopts this view.

[30] See also Kassler 2018, page xx.

[31] This view has been developed by Rogers 1979. See also Iliffe 2017, page 396.

[32] See Mintz 1962.

[33] Edwards 1699, pages 27–28.

any idea is innate."[34] Newton's views on morality, to be discussed more below, also did not rely on humans possessing innate ideas. Finally, it seems highly unlikely that Locke ever desired or planned to write a further work on innate ideas.

Materialism was commonly associated with Hobbes and Hobbesianism.[35] But it is harder to see why this view would strike at the heart of morality. Global materialsm, which denied God's existence, might do so. But Newton would have had to have been a very uncharitable reader of Locke to have read the *Essay* as advocating that view. A materialist ontology of human beings was often associated with a denial of the afterlife. And this might strike at the root of morality. But, as discussed below in connection with his mortalism, Newton well understood that the ontology of persons was a separate issue from life after death. And as with innate ideas, it is hard to imagine that Locke ever considered pursuing another book advocating materialism. Indeed, pursuing such a work would have been a surprising, and even dangerous, plan.

The suggestion that Newton accused Locke of striking at the root of morality and being a Hobbist on the basis of Locke's alleged materialism about human beings also faces another puzzle. Why, if Newton believed this was Locke's position and also believed that it was a truly wicked and impious one, would he move to patch up their relationship and remain friends with Locke? We know that Newton was particularly spiky toward those he saw as irreligious. According to Catherine Conduitt, "He could not bear to hear any one talk ludicrously of religion, often angry with Dr Halley on that score, & lessened his affection for {D^{r}} Bentley."[36] Conduitt also reports that Newton "broke off all acquaintance" with Vigani, who he had previously enjoyed conversing with, after the latter told "him a loose story about a Nun."[37] Why would the case be different with Locke? The question is particularly pressing just because Locke *did in fact* think materialism about human ontology was possibly true. He resolutely defended the epistemic and moral viability of the view up until his death. So, if Newton really did accuse Locke of Hobbism on these grounds, then his *volte-face* becomes somewhat hard to understand.

This brings us to the third possibility, that Newton was upset by something Locke had said about morality.[38] This suggestion seems, to me, to be eminently plausible. Locke was committed to a rather stark version of hedonism on which good was to be identified with pleasure and evil was to be identified with pain. This is clearly something that caught Newton's eye. Among the pages he dog-eared in his copy of the *Essay* was one with this passage:

[34] Quoted at Koyré 1965, page 282. For discussion, see Rogers 1979. See also the reports of Conti translated at Iliffe 2006, volume 1, page 241.

[35] For helpful discussion see Duncan 2022.

[36] Keynes Ms. 130.07, 1v.

[37] Keynes Ms. 130.07, 1r.

[38] For Dempsey's brief consideration and rejection of this view, see Dempsey 2017, page 456.

> Good and Evil, as has been shewed in another place, are nothing but Pleasure or Pain, or that which occasions or procures Pleasure or Pain to us. *Morally Good and Evil* then, is only the Conformity or Disagreement of our voluntary Actions to some Law, whereby Good or Evil is drawn on us, from the Will and Power of the Law-maker; which Good and Evil, Pleasure or Pain, attending our observance, or breach of the Law, By the Decree of the Law-maker, is that we call *Reward* and *Punishment*.[39]

The Hobbesian tone of this passage hardly needs elaborating. In Newton's mind it also might have been associated with Epicureanism, a set of philosophical views we saw his stark opposition to in Chapter 2. It is also likely that these definitions would have sat uneasily with Newton's conception of morality. Newton seems to have been a proponent of what we would now call divine command theory. For him, the commands issued by God in the Old Testament were the root of morality.[40] Ultimately, of course, Locke did accept some version of divinely orchestrated natural law theory alongside his hedonism. But Newton would hardly be alone in wondering how these two aspects of Locke's moral thought were meant to be consistent. The issue continues to be a mainstay in secondary literature devoted to Locke's moral thought.[41]

Locke's hedonism about value alone might have been enough to closely associate him with Hobbes in Newton's mind. But this "Hobbesian" theory of value matched with Locke's views on the will might have made the association even tighter. In the first edition of the *Essay*, the only one published by the time of the letter in question, Locke had adopted a seemingly deterministic theory of volition on which the will was simply determined by the greatest apparent good. As seen in the previous chapter, Newton had strong views about the will and saw freedom of will as central to moral accountability. So, again, we have a plausible reading of Locke on which he held a view associated with Hobbes that Newton might have seen as a threat to morality and right conduct.[42]

What about Newton's claim that Locke was planning to pursue his offending views in another book? Here as well I think we have a potential match. William Molyneux asked Locke on a number of occasions, beginning in 1692, to produce "*a Treatise of Morals*, drawn up according to the Hints you frequently give in Your

[39] Locke 1690, page 157. See *NL* #967. Rogers 1979, pages 199–200 and Iliffe 2017, page 396 also note Newton's dog-earing of this page in their discussions of this issue.

[40] See, for instance, Newton's annotations to the end of the Queries in his 1717 edition of the *Opticks*, reproduced as illustration 10 in Manuel 1963.

[41] See Rossiter 2016 for a helpful point of entry.

[42] Intriguingly, by the time of Newton's September 1693 letter of apology, Locke had already been at work on revisions to his account of the will, revisions which appeared in the second edition of the *Essay*. In July of that year Locke shared with Molyneux a new outline of the chapter which introduced his views on suspension of the will. See Locke 1976–1989, volume 4, page 701. These revised views might have been more amenable to Newton. The second edition of the *Essay* was also, of course, the first to contain Locke's famous discussion of personal identity.

Essay."[43] And Locke agreed to make an effort in this direction.[44] The short essay known as "Ethica A," which dates from this period, might even be the product of this effort.[45] So, it is possible Locke had mentioned to Newton his intentions or Newton had gotten wind of them in some other way.

Again, Newton's state of mind in the summer and fall of 1693 gives us some reason for being cautious about his letter to Locke and the understanding of the *Essay* that gave rise to it. That being said, the most plausible interpretation is that Newton had been alarmed by Locke's views on hedonism and the will, rather than by any of Locke's claims about the ontology of human beings. And certainly by the time of the Gregory memorandum in 1694, by which time the second edition of the *Essay* containing Locke's view on personal identity had gone to press, Newton had views which would have made him amenable to Locke's treatment of these issues. All told, I believe that the analysis offered in the previous section can stand.

8.5. Mortalism

Christian mortalism is a heretical position according to which the human soul is not naturally immortal. Instead, following the death of the body, the soul depends on a special dispensation from God to continue in existence. Some mortalists, so-called pyschopannychists, believe that the soul continues on in existence but is asleep prior to the Resurrection. A more radical view, called thnetopyschism, holds that the soul literally ceases to exist at death and is recreated at the Day of Judgment. Mortalism gained a number of converts in seventeenth-century England, as is perhaps best exemplified by the popularity of Richard Overton's *Mans Mortalitie*.[46] And many of Newton's closest confidants, such as John Locke, were mortalists or had clear mortalist sympathies.[47] There is a general consensus that Newton adhered to some form of mortalism.[48] As he writes at one point: "Human souls are immortal, not through natural causes but through the will of God."[49] But there are questions about what this fact means for his views on the ontology of minds.

Dempsey has argued that Newton's commitment to mortalism provides evidence for his commitment to the claim that consciousness is a necessarily embodied phenomenon. The suggestion is that Newton's adherence to the latter view

[43] See Locke 1976–1989, volume 4, page 508. See also pages 533 and 602.

[44] See Locke 1976–1989, volume 4, page 524.

[45] Locke 1997, pages 318–319.

[46] For background to Christian mortalism and its popularity in early modern Britain, see Burns 1972 and Ball 2008.

[47] For Locke's mortalism, see Lucci 2021, ch. 4.

[48] See Force 1994 and Dempsey 2006, sec. 3. See also Force 1999 and Smolinski 1999. There is an open question about whether Newton was a psychopannyschist or a thnetopyschist. For some relevant considerations and positions taken in the literature, see Snobelen 2005, pages 263–264; Dempsey 2006, footnote 24; Dempsey 2011, page 255; Dempsey 2017, sec. 4.

[49] CUL Ms. Add 3965, 547r.

might be a consequence of his adherence to the former one.[50] And this, in turns feeds into the larger monist interpretation developed by Dempsey and discussed in the previous chapter. But as Gorham has pointed out, and Dempsey himself acknowledges in several places, there is no necessary connection between any given view on the ontology of the mind and its (im)mortality.[51] Further, we should note that Newton endorsed mortalism on primarily Biblical grounds.[52] And we should recall his injunction against overly metaphysical interpretations of Biblical teachings. So, while Newton certainly believed in the natural mortality of the soul, it is not obvious that he thought this entailed a form of monism or a rejection of immaterial substances.

There is one text considered by Dempsey that might seem to run contrary to this thought, but even here things are not so straightforward. The relevant passage comes from Newton's "The Question Stated about Absteining from Blood."[53] Dempsey, following Snobelen, claims that Newton identifies *nephesh* (נפש, commonly rendered as *anima* or life) with the blood of animals.[54] This would mean Newton took the *anima* or life itself to be corporeal and thereby rejected immaterial human minds.[55] But there are a few things to note. First, the passage is explicitly concerned with the life of non-human animals. So, we should be cautious in drawing conclusions about the ontology of humans from Newton's claims in it. Second, at issue in the text is a word, *nephesh*, used in a law given to Noah and referenced by Moses. As Newton notes, "In six or eight hundred years, languages alter very much," meaning it was incumbent on Moses to offer a gloss or interpretation of what the word might mean.[56] So far so good, and Newton does indicate that Moses identifies the *nephesh* with the blood as a way of helping the Israelites understand the command. But the most salient portion of the text comes after this when Newton writes the following: "This word here must signify a corporeal substance which can be eaten & therefore not the life but the blood in which the life is seated or the blood of the life as it is called in the next words Gen IX. 5."[57] Because the law

[50] Dempsey 2006 and Dempsey 2014, sec. 3.

[51] Gorham 2011b, note 26; Dempsey and Stoyles 2010; Dempsey 2011, sec. 4.

[52] In connection with this, it is worth mentioning that Newton's reading notes on Ralph Cudworth's *True Intellectual System* suggest he understood belief in the immortality of the soul to have a non-Biblical origin: "The Ægyptians were the first assertors of the souls immortality & of its Transmigration after death." *Out of Cudworth*, UCLA Clark Library, Annex 3, Misc. Mss and Broadsheets Box 4, Folder 10, 1v. In connection with this see *NCS* #446 at page 336. Several other early modern mortalists were at pains to stress the pagan origins of belief in the immortality of the soul. See Ball 2008, pages 49–50.

[53] This manuscript is now in private hands, but was sold as part of Sotheby's Lot 232. A transcription is available from the Newton Project.

[54] Snobelen 2005, page 264.

[55] See Dempsey 2006, page 430 and Dempsey 2011, page 243.

[56] In connection with this, it is interesting to note that Newton dog-eared a page in his copy of Locke's *Essay* where the discussion related to words changing their meaning over time. See Locke 1690, page 133. *NL* #967.

[57] "The Question Stated about Absteining from Blood," 2r, underlining added.

is about not eating something, and because we can only eat corporeal things, Newton seems to infer that *nephesh* refers to the blood and *not* to the life of the animal. And we ought to refrain from eating the blood because the life—something distinct from the blood—is in the blood. But in order for the sense of the passage to work it must be that Newton is thinking of life as something incorporeal. Put differently, Dempsey is right that Newton identifies *nephesh* with blood, but wrong that Newton identifies *nephesh* with life, soul, or spirit. At least, this seems to me to be the best interpretation of this Newtonian text. I freely admit the passage is a difficult one and that other interpretations are possible. But I would also add that Newton himself thought a great deal of caution was called for when trying to understand Biblical Hebrew terms; as he wrote to Caspar Neumann: "The Hebrew tongue is said to be narrow, & the just signification of several words to be almost lost & few books are extant written in the ancient Hebrew, & the designe of recovering the ancient signification of the words must be very commendable" before noting that he himself was not knowledgeable enough to undertake the task of recovery.[58] At a minimum, it seems to me that Dempsey's interpretation is not obviously correct and is in need of further articulation and defense. So, even if this text shows Newton breaking his own rules about reading ontology into or off of Scripture, it is far from clear that it shows him finding a monist ontology of humans in the Biblical text.

My suggestion, therefore, is that Newton's mortalism might not have obvious consequences for how he understood issues relating to monism or dualism or the (in)corporeality of the soul. Again, recalling that Newton accepted the view on largely Biblical grounds can be helpful. Newton rejected metaphysical readings of Scripture in part because he worried they would obscure the clear *moral* lessons to be drawn from Scripture. And the moral implications of mortalism are clear enough. The view makes human existence after natural death an unwarranted and special dispensation of God. Put differently, the view emphasizes God's ultimate and magisterial dominion over humans: "by the same power by which God gave life at first to every species of animals he is able to revive the dead."[59] Recognizing that not only our existence in this life but also our potential existence in the afterlife depends on God in this way should only serve to increase our subservience and incentivize our obedience to God's commands.

I now want to suggest that keeping this point about God's dominion in mind can help us to see an important link between Newton's mortalism and his metaphysical commitments, specifically his understanding of minds as secondary substances. Recall that secondary substances are nothing over and above collections

[58] *NCS* #X.960. On Newton's limited knowledge of Hebrew, see Joalland and Mandelbrote 2016.
[59] Keynes Ms. 9, 2r. See also the repetition of the claim at 3r and 4r.

of powers. They have no essence, nature, or substratum capable of securing or explaining their existence. This ensures their fundamental dependence on God. Compare this with James Force's characterization of Newton's mortalism: "For Newton, there is nothing intrinsic to the human soul which empowers it to survive the death of the body as the result of its own innately immortal qualities unless God wills it."[60] It seems to me there is a very clean fit here. Indeed, Newton's account of secondary substances makes the idea of something created yet naturally immortal, that is to say capable of continued existence on its own, a contradiction or a very basic confusion. Anything that is created by God must be such that it fundamentally depends on God.[61] So, on this point, there is a deep and important affinity between Newton's theological commitments and my reconstruction of his metaphysical ones.

Here is a slightly different twist on the same basic point. Newton's view was that the message of Scripture was fundamentally moral rather than metaphysical. God's ownership and sovereignty was analogous to that of a monarch. In the early modern period, monarchical control could be conceptualized with regard to the monarch's power of life and death over her subjects. Thus, it was commonly believed that sovereigns were justified in enforcing capital punishment—Newton's time at the Mint gives ample testimony that he shared this view of the sovereign's power. On this line of thought, part of what is required in order for God to have monarchical control over humans is for God to have power over our life and death. And God's moral sovereignty would therefore sit uneasily with the idea that souls are, by their very nature, immortal.

The point here is not that Newton sought to extract his metaphysics of human minds directly from what he took to be Scriptural teachings about mortalism. Rather, the point is that Newton had mortalist commitments derived from Scripture and that those commitments informed and influenced the metaphysical picture he developed. James Force has argued that "God's dominion is the fundamental first metaphysical principle underlying Newtonian metaphysics" and that recognizing this can help us to unite various elements of his thought.[62] While I believe Force's view of Newton is overly synoptic, it certainly seems as though this is one instance in which recognizing Newton's emphasis on God's dominion can help us to see a deep affinity between his theological mortalism and his ontology of substances.

[60] Force 1994, page 184. And just before: "For Christian Mortalists, there is nothing inherent in the human soul which necessitates its spiritual existence after the death of the body *if* God does not act to reawaken or re-create it."

[61] In this sense, I disagree with Force's claim that Newton allows for the possibility that human souls might have an "inherent immortal existence" provided the soul was unconscious whenever separated from the body. See Force 1994, page 194.

[62] Force 1990a, page 84. See also Force 1994, page 180.

8.6. Rejecting the Agnostic Interpretation

Up to this point, this chapter has argued that Newton had available to him a well-motivated form of agnosticism about the ontology of human beings. Both dualism and monism were compatible with his basic metaphysical and theological commitments. Ultimately, however, my view is that there is simply not enough available evidence to claim that Newton ever took this agnostic view seriously and he should not be interpreted as having endorsed it. Instead, he seems to have remained a committed dualist.

While some of Newton's language in Query 28 is rather cagey in a way that might point to agnosticism, there are other late texts that look decidedly more dualist. For example, in drafts for the Queries, Newton at various places considers an electrical spirit and various phenomena it might be responsible for. One of these is "uniting the thinking soul & unthinking body."[63] The same language appears in another version of the claim.[64] This, of course, indicates that the soul and body are distinct and stand in need of something to facilitate their union. And, ultimately, it is this steadfast commitment to a certain understanding of mind–body interactionism that supports the claim that Newton remained a dualist throughout his career.

The basic picture is as follows. Human bodies are physical objects. Almost all of their activities are the product of purely physical (albeit not entirely mechanical) processes. This is true both for processes we would consider to be passive and those we would consider to be active. So, for example, in passive cases of sensation (understood here at a high level of abstraction), some body collides with the eye, which produces motion in the fluid inside of the eyeball, which produces motion on the retina, in turn moving the animal spirits in the optical nerve, and ultimately producing some motion in the brain.[65] And, for example, in active cases of moving an arm (again, understood at a high level of abstraction), some motion in the animal spirits near the brain produces a chain of motions ultimately terminating in the kinds of contractions or dilations of muscles that cause the arm to move.[66]

Ultimately, in both of these cases there will be something non-physical. At the very end of the passive process of sensation, the motion produced through physical processes in a very particular part of the brain will cause the mind to have a perception. And at the very beginning of the active process of bodily motion, the mind through an act of willing will produce the initial motions of animal spirits that eventually result in the motion of the arm.

Consider the difference between a Cartesian ontology of human beings and a Scholastic one. For the Scholastics, the substantial form is wholly present in the

[63] CUL Ms. Add 3970.3, 235r.
[64] CUL Ms. Add 3970.3, 241r.
[65] For more details concerning the relevant physiology, see Hamou 2014 and Kassler 2018.
[66] For more details concerning the relevant physiology, see Wallace 2003 and Kassler 2018.

entirety of the human body. It animates the body, as it were, from every place in the body. By contrast, for Descartes, the pineal gland is the only part of the body which is actually in "contact" with the soul. The rest of the body's processes are mechanical. The basic picture sketched above—while importantly different from Descartes'—is much more like the Cartesian view than like the Scholastic one.

There are two important senses in which the view being outlined here is dualist rather than monist. First, this view "confines" the soul to a certain very restricted part of the body. This is important because doing this will make for a natural and principled distinction between that part of the body—the part that directly interacts with the soul—and the rest of the human body, which is purely physical. This will make it hard to claim that the patella, for example, is really part of the same substance as the soul. Second, and relatedly, this view is *interactionist*. It seems to presuppose that there are types of processes that are different in kind; that, at a certain stage, the physical story will run out and new explanations must be invoked. And this, of course, feeds into the idea that there are two different things producing changes in one another.

Call the basic view sketched above the sensorium account of mind–body interaction. Newton was committed to this view from very early on. In the previous chapter, we saw relevant texts from "Of the Soul" in *CPQ* where Newton makes clear that the soul is only able to "see" certain parts of the brain. This text also contains a list drawn from Henry More of potential candidates for the part of the body that is the sensorium.[67] We have also seen the sensorium account on full display in *De gravitatione*. Recall Newton's claim that while the power of mind–interactivity is possessed by all bodies, it is only triggered in some of them and only on certain occasions. There he wrote about "the fact that the parts of the brain, especially the more subtle ones to which the mind is united, are in continual flux."[68] So again, it is just in the sensorium that this interaction happens.

The same account appears in Newton's early optical writings. He writes that:

> [T]he ends of the Capillamenta of the optique nerve, wch pave or face the Retina, being such refracting Superficies, when the rayes impinge upon them, they must there excite these vibrations, wch vibrations (like those of Sound in a trunk or trumpet,) will run along the aqueous pores or Crystalline pith of the Capillamenta through the optic Nerves into the sensorium (which Light itself cannot doe,) & there I suppose, affect the sense with various colours according to their bignesse & mixture.[69]

[67] *CPQ* 383.
[68] *DG* 49.
[69] *NCS* #146.

Again, the idea is that the mind only perceives colors once motion is transferred to the sensorium. The same thought is clearly present in some text Newton prepared for a planned reprinting of the *New Theory About Light and Colours*. Newton there writes of the "kinds of those actions (suppose whether they be pressions, strokes, or other dashings) by which light may produce in our minds the phantasms of colours" and claims that "I do not myself esteem colours the qualities of light, or of anything else without us, but modes of sensation excited in our minds by light."[70] And the model seems to be in operation in the earlier "Of Colours."[71] So the sensorium account was well-established in Newton's thought during the early parts of his career.

My view is that this basic commitment to the sensorium account—a fundamentally dualist account—continues up through Newton's death. And this means that he never came to reflectively endorse the kind of well-motivated and appealing agnosticism that was open to him and that may have better cohered with his considered epistemic commitments. One late text which seems to show Newton continuing to endorse the sensorium account is the Queries to the *Opticks*. Consider even Query 28, which above was used to motivate the possibility of agnosticism in Newton's thought:

> Is not the Sensory of Animals that place to which the sensitive substance is present, and into which the sensible Species of Things are carried through the Nerves and Brain, that there they may be perceived by their immediate presence to that Substance?[72]

Newton does shy away from language of souls or minds here, but the model he is invoking is clear enough. Physical processes produce motion in a certain part of the brain. At a certain stage, the physical process stops and something different happens, a non-physical object has a perception. Query 24 offers a version of the sensorium account for agency and active processes in the body initiated by the soul.

Perhaps nowhere is this commitment to the sensorium account more pronounced than in Newton's claim that space is (as it were) God's sensorium. This is a commitment Newton maintained throughout the late stages of his career up until his death, despite considerable pressure to abandon or reformulate it. Correctly understood, it is also a claim that supports an understanding of the sensorium account as a dualist one. And this, in turn, supports the idea that Newton remained committed to a dualist ontology of human beings.

70 Transcribed at Cohen 1958, page 365.
71 CUL Ms. Add 3975, 19–20.
72 *Opticks* 344–345.

8.7. God's Sensorium

Newton's claims that space is the divine sensorium first appeared in the 1706 edition of the *Optice*. Here are the relevant passages as they appeared in the 1730 version of the *Opticks*:

> [D]oes it not appear from Phænomena that there is a Being incorporeal, living, intelligent, omnipresent, who in infinite Space, as it were in his Sensory, sees the things themselves intimately, and thoroughly perceives them, and comprehends them wholly by their immediate presence to himself: Of which the Images only carried through the Organs of Sense into our little Sensoriums, are there seen and beheld by that which in us perceives and thinks.[73]

> [T]he Wisdom and Skill of a powerful ever-living Agent, who being in all Places, is more able by his Will to move the Bodies within his boundless uniform Sensorium, and thereby to form and reform the parts of the Universe, than we are by our Will to move the Parts of our own Bodies.[74]

These passages are well-known and have engendered considerable debate and confusion among Newton's readers. Leibniz, for instance, reports that he found them so utterly bizarre and problematic that, upon first reading them, he laughed.[75]

In an earlier paper, I attempted to situate Newton's claims within his larger philosophical thought, defend their cogency and orthodoxy, and push back against readings that made them merely figurative or metaphorical.[76] I believe that in this instance Newton meant—more or less—what he wrote. A few points from this previous paper will be particularly helpful for our current purposes.

Leibniz found a number of different ways to put pressure on Newton's claim that space might be like a sensorium for God. One of these involved insisting that sensoria were corporeal entities, they were parts of the brain. Therefore, if God were to have a sensorium, then God would also have a body. Leibniz also insisted that if God were to have a sensorium then it would be a part of God, something that helped to compose God. But God is meant to be a perfectly simple and indivisible being. My view is that these accusations involve a fundamental misunderstanding of Newton's claim.

[73] *Opticks* 345. A note on the "as it were" in this passage. As Koyré and Cohen 1961 notes, the inclusion of the Latin word *tanquam* in the 1706 *Optice* is the result of a cancel: a page reprinted and pasted into the text after the rest of the book had been printed. A small number of copies of the original printing—missing the *tanquam*—were sold. Dhondt and Ducheyne 2021, sec. 3 shed some further light on the issue. See also Connolly 2014, sec. 13.

[74] *Opticks* 379.

[75] *NCS* #1138.

[76] See Connolly 2014. This view is considered and criticized in Henry and McGuire 2018, Kassler 2018, and Henry 2020b.

Leibniz's charge rests on understanding Newton to be saying that God's relationship to space is like the relationship of a person's *body* to their sensorium. A person's body has the sensorium as a proper part. So we can certainly talk about, say, Fatima's sensorium in a bodily register. By this, we mean to pick out a part of the brain that is a proper part of her body. If Newton had been meaning to channel this way of talking when claiming that space was God's sensorium, then he clearly would have committed himself to God having a body and space being one of its constitutive parts.

But Newton is better read as claiming that God's relationship to space is like the relationship a person's *mind* or *soul* has to their physical sensorium. So, it is possible to talk about Fatima's sensorium with reference to her body. But it is also possible to talk about Fatima's sensorium and mean to pick out the physical object that bears a special relationship to Fatima's soul. In this sense, we can say that Fatima's soul "has" a sensorium. There is a particular body to which it bears a special relationship; it is not a disembodied spirit. Importantly, in claiming that her soul has a sensorium, we are not saying that the sensorium is part of her soul or that her soul is bodily in virtue of having the sensorium. I have a Toyota Prius; I stand in a special relationship to that particular car. But this does not make me an automobile and the Prius is not one of my constituent parts.

I take it that Newton is trying to warn us away from Leibniz's interpretation and trying to push us more toward an interpretation like mine in this passage:

> [W]e are not to consider the World as the Body of God, or the several Parts thereof, as the Parts of God. He is an uniform Being, void of Organs, Members or Parts, and they are his Creatures subordinate to him, and subservient to his Will; and he is no more the Soul of them than the Soul of Man is the Soul of the Species of Things carried through the Organs of Sense into the place of its Sensation.[77]

The first part of the passage warns us against thinking about space as God's sensorium in the way we think of Fatima's sensorium as part of her body. And the latter makes clear that the relevant relationship is the one between the sensorium and the soul, not the sensorium and the rest of the body.

In order for this reading to work, however, it must be the case that Newton is adopting a dualist approach to human beings. The soul has to be something distinct from the body it interacts with. If human souls and bodies were united into one single substance, then claiming that God's relationship to space was like a person's relationship to their sensorium *would* entail that God was corporeal and had parts. So, properly understood, Newton's claims about the divine sensorium

[77] *Opticks* 379.

support the claim that he remained a dualist even through the later stages of his career.

8.8. Souls and Sensoria

Jamie Kassler has published what is, to date, the most in-depth study of Newton's claims regarding human and divine sensoria.[78] In part through careful attention to Newton's texts and in part through a rigorous examination of the views of Newton's contemporaries Thomas Willis and William Briggs, Kassler has added a wealth of detail to our understanding of Newton's views on the anatomy and physiology of perception and voluntary action. My focus, of course, is on Newton's fundamental metaphysics. So, I have little to add to Kassler's interpretation of Newton on the mechanics of bodily actions and passions. There are, however, two aspects of Kassler's interpretation that appear to clash with claims made above and therefore deserve some closer examination.

Recall the passages from Query 28 examined at the outset of this chapter. I had considered the idea that Newton's reference to a "sensitive substance" might reflect a shift toward a more agnostic attitude but ultimately concluded that even these passages reflect a dualist approach. By contrast, Kassler believes that the sensitive substance mentioned in this text is something distinct from both the human soul and from the other parts of the corporeal perceptual apparatus. On her view, it refers to a physical anatomical substructure which is the material cause of our perceiving and (intentionally) acting. This, then, leads Kassler to claim that there is a contradiction between Newton's claim from Query 28—where humans are said to "perceive things indirectly, that is, by a mediator or 'third' thing" (i.e. by means of this sensitive substance)—and Newton's claim in Query 31 where both God and the human soul are said to perceive things directly.[79]

I believe there is no such contradiction between Queries 28 and 31. Instead, Kassler can be seen as making the same mistake as Leibniz. Both human souls and God perceive what they are immediately present to. God is immediately present to all objects. Human souls, by contrast, are only present to certain physical objects or processes in a specific part of the brain. This means that the relationship that God has to all of space is like the relationship that the human soul has to one very specific part of space. So God and human souls both perceive (some) things directly. Ultimately, Kassler's claim that *humans* perceive things indirectly will be true. Human perception involves layers of physical processing between external objects and our awareness of those objects. So the soul does not perceive *external*

[78] Kassler 2018. For helpful discussion of and some critical engagement with Kassler's book, see Vida 2019.

[79] Kassler 2018, page 27.

objects directly, it learns about them indirectly insofar as they are the distal causes of the physical objects or processes present in the sensorium. But this kind of indirectness is wholly irrelevant to the matter at hand. Newton can—and does--still claim that the soul directly perceives those physical objects or processes present in the sensorium. This is the similarity that undergirds his claim that space is (as it were) a divine sensorium. Again, Newton is claiming that God is to space as the soul is to the sensorium. Newton is not claiming that God is to space as human beings are or the human body is to the sensorium.[80]

Ultimately, I believe there is no need to reify the "sensitive substance" mentioned in Query 28 in the way that Kassler does. What Newton refers to there is just the soul. Fortunately, even if I am right about this fairly minor textual issue, there will not be much damage done to Kassler's interpretation. Her project of isolating and understanding various components of our neural anatomy that Newton saw as particularly salient for perceptual and volitional processes is certainly a salutary one. Whether or not we refer to these components as a "sensitive substance" (or *anima*, in Kassler's parlance) is purely a matter of semantics. And very little hangs on the question of whether (or the extent to which) these anatomical components are genuinely distinct from the rest of the body in a way that would allow them to be classified as a "third thing." I would just resist the claim that the "sensitive Substance" referred to in Query 28 is a physical entity, rather than the "immaterial" soul (or *animus*, to use Kassler's term). I would also insist that whatever physical processes these neural components undergo, they certainly do not "perceive" or "sense" in the sense of having mental states or phenomenology. This is also something reserved for the soul itself.

Second, Kassler claims, in part drawing on Newton's diagram of his reflecting telescope, that the human soul "must be seated *above* the coat of the *sensorium*, so that its intellectual eye, by looking downwards from some undisclosed place in the anatomical space of the brain, beholds the spectacle represented to it . . . at the anterior of the under surface of the sensorium."[81] This understanding of human sensoria, when applied to God through the analogy of nature, funds the following understanding of space as the *sensorium dei*:

> [T]he unbounded vacuum beyond the material cosmos is the place of the divine spectator, whose all-seeing intellectual eye beholds directly, that is to say, looks down upon the moving spectacle in the region of cosmic space. [¶] And recalling that intellect, the rational knowing power of *animus*, perceives, as well as thinks, and that the space beyond the material cosmos is a vacuum, then it is also plausible to assume that . . . First, in that part of the vacuum, which commences where

[80] Shein 2012, pages 30–31, offers an admirably clear statement of the view I mean to be advocating for here.

[81] Kassler 2018, page 102.

> the sphere of the fixed stars ends, God perceives 'things' directly, because his all-seeing eye of intellect surrounds and also 'comprehends' (i.e. takes in, contains) the material cosmos (perhaps like the pupil in the physical eye is surrounded by the iris). Second, in the unending space beyond the intellectual eye, God's 'thinking substance'. . . forms intellectual plans (design as model) and intellectual purposes (design as intention.)[82]

I have grave reservations about this interpretation.[83] These can be put under the twin headings of ubiquity and uniformity.

First, I worry about the degree to which this position will align with Newton's understanding of divine omnipresence.[84] In a clause elided from the block quotation, Kassler offers her interpretation as providing insight into Newton's repeated insistence that the divine sensorium occupies the whole of space. But, just as the iris surrounds the pupil rather than being present in it, it is somewhat difficult to see how God will, on this picture, truly be present *in* rather than surround the cosmic space. Suppose that on a cloudy day I commit a sin on the bottom story of a below-ground parking garage. How will God, looking down from the edge of the fixed stars be able to perceive what I have done? On my understanding, to say that space is God's sensorium is to say that God bears the same relationship to *every part of space* that our souls bear to the part of space occupied by our sensorium. God does not "look down on" these regions. God is present "within" them. God does not look down on my sins from the fixed stars, God is co-located with me in the parking garage and is aware of my sins (at least in part) by being present to them. As Newton writes: "We are to consider him as invisibly present to us at all times & in all places & that he knows all that we say or do: for in him we live & move & have our being."[85] This is what abets God's omniscience and monarchical control.

Second, this proposal seems to align different regions of space with different parts of God. The "intellectual eye" is present as a hollow orb surrounding the cosmos whereas the "thinking substance" is present in the empty spaces beyond that hollow orb. But, as seen in Chapter 3, Newton consistently affirms God's simplicity and uniformity. As the General Scholium puts the point: "all of him is like himself: he is all eye, all ear, all brain, all arm, all force of sensing, of understanding, and of acting."[86] Or as one of his manuscripts has it: "the father is a pure spirit . . . being alike in all places."[87] The worry is that Kassler's interpretation will

[82] Kassler 2018, page 114–115. See also page 132.
[83] Levitin 2022, page 688, also raises worries for this portion of Kassler's position, though on different grounds than those I offer below.
[84] A similar point is made at Vida 2019, page 146.
[85] Keynes Ms. 3, 35. See also CUL Ms. Add 3968, 414v and 436r.
[86] *Principia* 942. Intriguingly, Newton uses very similar language in a manuscript where he is explicitly responding to critiques of his claims about the divine sensorium. See CUL Ms. Add 3968, 41:23r.
[87] Yahuda Ms. 15.3, 67r.

give ammunition to the Leibnizian complaint that if God has a sensorium God will also have distinct parts.

8.9. Conclusion

This chapter has sketched out a Newtonian path not taken. Specifically, it has argued that core components of Newton's metaphysics made possible a monist ontology of human beings. Newton might been agnostic about whether human beings were one substance or whether they were a combination of two distinct substances. Such a view might even have been more appropriate for him to hold given core components of his epistemology. Further, Newton's relationship with Locke would seem to have put him in a good position to understand this view and its attractions. Finally, such a view, while it would not have been necessitated by Newton's mortalism, may have nicely cohered with it. Ultimately, however, Newton never seems to have adopted this agnostic approach to human beings. He remained wedded to a dualist interactionist view. This comes through in many texts, but is perhaps best seen in the various passages where he attempts to shed light on God's relationship to space by discussing the ways in which human souls relate to sensoria in the brain.

9
Conclusion

The chapters in the body of this book have worked to chart the contours of Newton's metaphysical system. Much of this work has involved contextualizing Newton. I believe it is important to remember, as Dobbs writes, that "Newton looked forward in time but backward as well."[1] Situating Newton against his predecessors and contemporaries has, I hope, proved helpful. Much of the book has also taken the form of argument. Some arguments offered have related to the logical space open to Newton and the advantages or disadvantages of various positions. Others have been textual and interpretative. In particular, I have drawn heavily on material from *De gravitatione*. Still other arguments have related to the secondary literature. I have worked to understand and consider the interpretations put forward by other commentators and to accept, reject, or amend their views as I saw best. All of this has been necessary to develop and motivate my global interpretation of Newton's metaphysics of substance. But at points this detailed contextual and argumentative work may have obscured the interpretation itself. Here, then, I thought it might be useful to offer a bare characterization of Newton's core metaphysical views as I understand them: the picture, as it were, without the argument.

There is one, all-powerful, immaterial, infinite God. This God exists in, but is ontologically distinct from, space and time. God's existence in all of space is the manifestation of God's omnipresence. God's existence at each moment of time is the manifestation of God's everlastingness. God created all other things and has total and rightful moral dominion over them. Although God is insensible by humans, humans can nonetheless learn about God. Revelation is a key source of this knowledge. But there are natural sources as well. Most particularly, features of the natural world (particularly the complex yet ordered structures of biological organisms and the solar system) allow us to infer that God exists, is wise, and has providential concern for creatures. The fact that humans must learn about God through God's works, rather than through some innate idea of God or through some form of direct access, entails that humans are ignorant of God's essence. Accordingly, Newton rejects the ontological argument. Ignorance of God's essence also leads to puzzles surrounding the relationship between the divine essence and the divine attributes, specifically puzzles about how to reconcile the possession of multiple

[1] Dobbs 1991, page 5.

Newton's Metaphysics of Substance. Patrick J. Connolly, Oxford University Press. © Patrick J. Connolly 2025.
DOI: 10.1093/9780191959240.003.0009

seemingly heterogenous attributes with divine simplicity. Ultimately, however, Newton is not particularly concerned by these metaphysical puzzles surrounding divine ontology. This is because he thinks it more important to conceive of God in moral terms, as ruler or governor of the universe and of humankind.

The word "substance" is equivocal as it applies to God, on the one hand, and creatures, on the other. God has a substantial nature or essence. This nature or essence determines God's other features. There are two fundamental types of created substances: corporeal bodies and human minds. These created substances lack anything that plays the metaphysical role of an essence or nature. Instead, they are merely collections of powers co-located in space. These created (or, as I call them, secondary) substances are able to exist in virtue of nothing beyond God's initial act of creation and ongoing acts of conservation. Put differently, they do not need to inhere in or to be subsistent in a substratum or underlying substance. Secondary substances also enjoy considerable causal autonomy. Newton rejects occasionalism and concurrentism. In a way, this is fitting. Secondary substances just *are* powers; so action (or, at least, the capacity for action) is built into them. This has the, perhaps surprising, consequence that Newton understood matter to be active. It also has the, less surprising, consequence that Newton sidesteps certain difficulties facing the view that humans have free will.

Newton's approach to bodies is fundamentally atomistic. And he rejects a material plenum. Fundamental corpuscles, while extended, are nonetheless metaphysically unified and their spatial parts are inseparable, even by God. This is a function of the fact that the bodies themselves just are powers with a certain determinate spatial shape and extent. The size and arrangement of a fundamental body's spatial parts cannot be modified while one and the same set of powers stays in existence. Newton initially claims that bodies are composed of (or, are collections of) three distinct powers: the power of solidity which ensures that bodies cannot co-locate, the power of mobility which ensures that bodies need not remain fixed in a single spatial location, and the power of mind-interactivity which ensures that bodies can produce ideas in human minds and be moved by human wills. With the exception of this third power (regarding which, more below) this initial picture is broadly "mechanistic."

This view develops as Newton pursues empirical research in natural philosophy. Because there are empirical phenomena that cannot be explained mechanically, there must be further powers or active principles. Newton also aims to offer a more precise characterization of the forces these powers exert using mathematical tools. While Newton is able to do this successfully in some cases (e.g. gravitation) he fails to do so successfully in other cases (e.g. magnetism and cohesion). While Newton often speaks of laws, they are not metaphysically basic components of his worldview.

Human minds, like bodies, are comprised of different co-located powers. These powers correspond to discrete cognitive faculties. Newton identifies five of these: perception (the ability to receive ideas on the basis of certain bodily actions), imagination, understanding, memory, and volition. Importantly, the presence of a faculty of understanding means that Newton rejects imagistic analyses of human thought. And an independent faculty of the will reflects the importance that Newton afforded to human freedom and also, perhaps, his view that humans were created in the divine image. Certain monist ontologies were open to Newton and he might have understood a human being as a single, unified type of substance. Indeed, adopting a monist view or a position of agnosticism on the question might have better reflected some of his epistemic commitments. Nonetheless, he remained a staunch dualist. The two types of created substance are differentiated not because they are fundamentally different types of things; they are both collections of powers. Their difference consists in the mere fact that they are sets of different types of powers.

This, on my reading, is Newton's metaphysics of substance. I hope to have made his views clear and coherent. In the introduction, I claimed that Newton's views on these issues were innovative, perceptive, and (by the standards of the period) compelling. I have tried to make the case for this throughout the book. But here it might be useful to single out several dimensions on which I think Newton's metaphysics deserves high marks.

First and foremost, I think the distinction between primary and secondary substances that I have highlighted in Newton's thought has much to commend in it. Obviously, the distinction is not original to Newton. Nonetheless, he gives it his own twist. He does this by adopting the view—distinctive in the period—that created substances are constituted by sets of powers. They are nothing over and above their powers. He abandons the idea that they might have some underlying substratum or some essence at their core. In my estimation, this Newtonian view comes with notable advantages.

One somewhat obvious advantage is that Newton is able to dispense with the philosophically problematic idea of substance as substratum or substance as (perhaps Lockean) real essence. The *epistemic* difficulties associated with such a view, particularly for those with broadly empiricist commitments, are familiar and well-known. Unlike Locke, at least as he is commonly understood, Newton can avoid the embarrassment of positing a "supposed, I know not what."[2] Equally important, however, is the fact that Newton's view lets him side-step certain *metaphysical* difficulties. He is not on the hook for explaining the nature of the widely-accepted but nonetheless quite puzzling inherence relation. Similarly, he does not have to cash out the idea that certain features or qualities of bodies "flow from" an essence.

Admittedly, Newton does allow that God has a substantial nature or essence. So, the divine case does present a version of the relevant epistemic and metaphysical

[2] Locke 1975, page 305.

difficulties. But in this case, I believe the concerns can either be blunted or turned into virtues. Here are four relevant considerations. First, it might be that we *should* tolerate a certain amount of mystery in the ontological architecture of the supreme deity. Given the nature and limitations of human cognitive and epistemic capacities, and given the majesty and infinitude of God, we might expect some mysteries to remain. Second, while these mysteries will remain, one virtue of Newton's approach is that they will be excluded from the natural world. We have a sensible path forward that will allow us to understand, in a thorough way, what created substances are. Put differently, this approach has the benefit of cordoning off the problem or isolating it within a restricted domain. Third, while I believe it makes sense to understand created substances as mere collections of powers, there are substantive philosophical worries surrounding ontologies that consist *solely* of powers.[3] The basic concern is that a universe full of nothing other than dispositions will fail to manifest a concrete reality, each constituent of such a world will rely on other things to bring itself into being (to activate or trigger it). What is needed, the objector insists, is something fully concrete which is capable of grounding the whole system of inter-related powers. What I characterized as Newton's non-deflationary approach to the divine substance can do just that. Fourth, we might find reason to embrace the idea that God has a substance (regardless of potential difficulties for the view) and that creatures do not because it preserves a significant and valuable distinction between God and creatures. This brings us to the next core advantage I see for Newton's approach to primary and secondary substances.

There is a deep honesty in Newton's claims about primary and secondary substances. He wants, understandably and—by standards of the period—commendably, to preserve a radical ontological difference between God and creatures. This approach lets him do that. And it does so in a way that neatly reinforces the idea that created beings are fundamentally dependent on God. For creatures, there is no "there" there beyond their powers. There is no genuinely independently existing being at their core. There are only the powers. And these only exist at God's good pleasure. This obviously allows Newton to make good on his radical voluntarism. And, at the same time, it reinforces what he sees as an important moral point: our very being—moment to moment—depends immediately on God and we ought to honor and obey God accordingly. As is perhaps now clear, I have a certain amount of sympathy with Newton's worry that others in the period are "only verbally" able to claim that bodies are "created and dependent."[4] Newton refuses to give mere lip service to what he sees as important truths. Accordingly, he develops a view on which there is something more to say about how and why

[3] I discuss this issue in connection with Newton's metaphysics in more detail in sec. 7 of Connolly 2024.

[4] *DG* 47.

secondary substances are both created and dependent. Newton is able to point out and articulate a stark difference between God and creatures.

So, Newton avoids claiming that "substance" is univocal and I have argued that—for him—this was a good thing. At the same time, he is also able to avoid going too far in the other direction. While "substance" is equivocal, Newton need not make the further (and, to my mind, much trickier) claim that "being" or "existence" is equivocal. This is important to him insofar as it will let him retain a meaningful and intelligible commitment to the *imago dei* thesis. On my view, Newton thought comparisons between God and humans were imperfect but were nonetheless deeply important.[5] So, while it was important to preserve difference and distance between God and creatures, it was also important to not make this difference and distance too great. Making existence equivocal might threaten to do this. Approaches that lean too far in the direction of analogy present similar risks.[6] Thus, I argue, Newton's distinction between primary and secondary substances represents a sensible and attractive *via media*.

Issues regarding divisibility and indivisibility gave rise to some of the most intractable disputes of the period. How could one marry a physics of discrete, extended, atomic particles with a conception of continuous spatial extension? Newton charts a very clean course through these choppy waters. He holds fast to the truth that any quantity of extension, however small, will itself be infinitely divisible. So, he avoids what seems like an obvious confusion on the part of Henry More that there might be "perfect Parvitudes"—entities that are extended but nonetheless so minimally extended that they cannot be split.[7] At the same time, he also avoids the myriad difficulties associated with the view that fundamental bodies are themselves infinitely divisible. Because his fundamental corpuscles, understood in line with the theory of bodies developed in *De gravitatione*, are made of discrete and differentiated, albeit spatially extended, powers rather than quantities of "matter," we can understand why they are not subject to spatial division. The powers themselves have determinate spatial extensions and distributions. Something with a different spatial extension and distribution would be a different power. So when asked why not even God can divide a fundamental particle, Newton can do more than pound the table and flatly assert that this would be impossible. He has access to a principled and coherent atomism.

Finally, I believe Newton's dualism also deserves a certain amount of praise. A great many of his contemporaries (whatever their position on global dualism, materialism, or idealism) were united by the fact that they defined mind and

[5] See Connolly 2014, sec. 9.
[6] For a nicely articulated account of these points, see Newlands 2024.
[7] More 1659, page 115.

matter oppositionally. On these views, the mental and the material occupy distinct and mutually exclusive realms. This, as is well-known, makes it tremendously hard to see how they might relate to or affect one another. Newton's powers-based ontology offers a single realm in which both can be housed. He does not have to fall back into a hylomorphism which had been rejected on independent grounds. And he does not have to argue that our everyday lived experience of mind–body interaction is merely illusory. As previously acknowledged, by allowing that minds have powers of volition and perception that allow them to, respectively, affect and be affected by bodies and that bodies have a power of mind-interactivity, Newton does, in a way, solve the difficulties by fiat. But even here there is something to like. Newton, quite sensibly, recognized that sometimes explanations will bottom out.[8] Surely God is capable of affecting both bodies and minds such that they behave in coordinated ways. There is no reason to think this is one of God's incommunicable attributes. And, so, there is no reason to think that God could not easily delegate this coordinative work through the creation of powers adequate to the task.

Of course, Newton's metaphysical system is not without certain problems and oversights. And, of course, it would have greatly benefitted from Newton's having spent more time articulating his views and fleshing out the details of his system. Even still, I want to insist, there is enormous value in what Newton did do and in what we can build out from what he did do. Newton offers us insights, arguments, and perspectives on metaphysics that are otherwise missing from or very difficult to find in the more familiar systems of authors like Descartes, Locke, Malebranche, or Berkeley.

In closing, I would like to briefly return to some of the positions and considerations canvased in the Introduction relating to the unity or otherwise of Newton's thought and his various intellectual endeavors. One helpful place to begin is with some remarks from a recent paper by Andrew Janiak. In what is, on the whole, an interesting and insightful essay, Janiak writes that: "He [Newton] had an especially vexed relationship with metaphysics . . . . His attempt to leave an old discipline behind was not entirely successful."[9] My sense is that Janiak's views here are widely shared, even among those interpreters not drawn to positivist readings of Newton.[10] But I believe that the quoted sentence, as it stands, is manifestly false. I have tried to show in this book that, if we give ourselves just a small amount of interpretative leeway, we can discover a Newton who had a perfectly straightforward, and even quite healthy, relationship with metaphysics. He was highly successful in sloughing off metaphysical approaches, concepts, or practices that he

[8] I discuss this point further in Connolly 2024.

[9] Janiak 2023, page 223.

[10] Compare, for instance, Janiak's thought with one from Frederick Coplestone: "the tendency of Newton's thought . . . was to continue the purification of physical science from metaphysics." Coplestone 1959, page 153.

regarded as antiquated or unproductive. But we need not conclude that he regarded metaphysics as an old discipline. His goal was not escape from metaphysical thinking. Quite the contrary, I have tried to show that, when circumscribed within a certain domain, Newton's metaphysical views were deeply inter-related, highly systematic, well-motivated, well-defended, and forthrightly held. He was at peace with metaphysics.

Why, then, are authors like Janiak so motivated to claim that Newton's relationship to metaphysics was problematic? Here is another quotation from the same paper which, I believe, can start to provide an answer: "Newton made great strides in transforming natural philosophy even while keeping one foot firmly planted in the metaphysical past."[11] Here, then, we begin to see where the problem lies. The issue is not with Newton's metaphysics considered in isolation or considered on its own merits. The difficulties arise when we seek to put his metaphysical views in conversation with his natural philosophical achievements.[12] This is the real topic of Janiak's paper and, it must be said, he tackles it admirably.

I am in full agreement with the idea that there are deep puzzles about the precise relationship between Newton's metaphysics and his natural philosophical theory and practice. But I struggle to see how this fact licenses the verdict that he had a problematic relationship with metaphysics. Suggesting that it does requires us to center natural philosophy in his thought and to demand that everything else to be answerable to it. I do not want to deny that Newton's work in mathematics and physics is more important than his work in, say, chymistry or Biblical chronology. To do so would be foolish. Instead, I just want to insist that if we are to take the historical Newton seriously, then we ought to give the various different elements of his thought their due and we ought to consider them on their own terms, not just in relation to the *Principia* and the *Opticks*. Put differently, I think it is correct that in some ways Newton's metaphysics and his natural philosophy are not in obvious or full alignment and so there are very good questions to ask about the relationship between the two. We all have reason to consider and address questions in this domain. At the same time, I want to insist, we might also have reason to sit with some of the discomfort here. After all, there is no guarantee that Newton's thought *was* fully unified or coherent. It might be that not all of the pieces fit together in the ways we hope or expect. And we should be careful not to stretch or deform any particular piece just for the sake of such fit.

As I wrote in the Introduction, I remain open-minded on many questions about unity and diversity in Newton's thought. But I think these questions are best asked

[11] Janiak 2023, page 224.

[12] Again, Coplestone provides a helpful reminder of how familiar this thought is: "It is clear that Newton was a philosopher as well as a mathematician and physicist. But it is not so clear how his metaphysics fits with his views of the nature and function of physical science.... It can hardly be maintained, I think, that there is a perfect harmony between his metaphysics and his more 'positivistic' pronouncements about the nature of science." Coplestone 1959, pages 155–156.

only once we have done the preliminary work of clearly mapping out his thought within a specific domain. This has been my goal. And my hope is that it enables further research into Newton's metaphysics considered in isolation, further research on the connections between Newton's metaphysical views and those of his predecessors, contemporaries, and successors, and also further research on the relationship between his metaphysics and other parts of his thought.

Works Cited

Abbreviations

CPQ—McGuire, J.E. and M. Tamny (eds.) 1983. *Certain Philosophical Questions: Newton's Trinity Notebook*. Cambridge: Cambridge University Press.

CSM—Descartes, R. 1984–1991. *The Philosophical Writings of Descartes*. J. Cottingham, R. Stoothoff, D. Murdoch, and A. Kenny (eds. and trans.) 3 volumes. Cambridge: Cambridge University Press. References provided by volume and page number.

DG—*De Gravitatione* in Newton, I. 2014. *Philosophical Writings*, 2nd edition. A. Janiak (ed.). New York: Cambridge University Press. 26–58.

NL—Harrison, J. 1978. *The Library of Isaac Newton*. Cambridge: Cambridge University Press. References provided by item number.

NCS—Newton, I. 1959–1977. *The Correspondence of Isaac Newton*. H.W. Turnbull et al. (eds.) 7 volumes. Cambridge: The Royal Society. References provided by letter number.

NPW—Newton, I. 2014. *Philosophical Writings*, 2nd edition. A. Janiak (ed.). New York: Cambridge University Press.

Opticks—Newton, I. 1730. *Opticks: Or, a Treatise of the Reflections, Refractions, Inflections and Colours of Light*. London.

Principia—Newton, I. 1999. *The Principia: Mathematical Principles of Natural Philosophy*. I.B. Cohen and A. Whitman (eds. and trans.). Berkeley: University of California Press.

Manuscripts

CUL—Cambridge University Library. Transcription of manuscripts available from The Newton Project: https://www.newtonproject.ox.ac.uk/Manuscript images available from Cambridge Digital Library: https://cudl.lib.cam.ac.uk/

Bodleian—Bodleian Library, University of Oxford.

Dibner—Dibner Library Manuscripts, Smithsonian Institution. Transcription of manuscripts available from The Chymistry of Isaac Newton: https://webapp1.dlib.indiana.edu/newton/ Manuscript images available from Smithsonian Libraries Digital Library: https://library.si.edu/digital-library

Gregory—Gregory Papers, Royal Society of London.

Keynes—Keynes Collection of the Papers of Sir Isaac Newton, King's College, University of Cambridge. Transcription of manuscripts available from The Newton Project: https://www.newtonproject.ox.ac.uk/

Lehigh—Lehigh University Special Collections. Manuscript images available from Lehigh Library Digital Collections: https://preserve.lehigh.edu/

TCL—Trinity College Library, University of Cambridge. Transcription of manuscript available from The Newton Project: https://www.newtonproject.ox.ac.uk/

UCLA Clark Library—William Andrews Clark Memorial Library, University of California, Los Angeles. Transcription of manuscripts available from The Newton Project: https://www.newtonproject.ox.ac.uk/ Manuscript images available from Calisphere: https://calisphere.org/

Yahuda—Yahuda Collection, National Library of Israel. Transcription of manuscripts available from The Newton Project: https://www.newtonproject.ox.ac.uk/ Manuscript images available from National Library of Israel: https://www.nli.org.il

Published Works

Ablondi, F. 2013. "Newtonian vs. Newtonian: Baxter and MacLaurin on the Inactivity of Matter," *Journal of Scottish Philosophy* 11(1): 15–23.

Anstey, P. 2000. *The Philosophy of Robert Boyle*. New York: Routledge.

Archer, J. and C.P. Ragland. 2022. "Descartes on Voluntary Action and Universal Conservation," in G.E. Ganssle (ed.) *Philosophical Essays on Divine Causation*. New York: Routledge. 102–121.

Ariew, R. 2022. "A Metaphysical Element in Descartes and the First Cartesians: Non-Univocal Predication," *The European Legacy* 27(3–4): 227–238.

Badaloni, N. 1968. *Antonio Conti: Un abate libero pensatore tra Newton e Voltaire*. Milan: Feltrinelli Editore.

Baillon, J.-F. 2009. "Two Eighteenth-Century Translators of Newton's *Opticks*: Pierre Coste and Jean-Paul Marat," *Enlightenment and Dissent* 25: 1–28.

Baker, J.T. 1930. *An Historical and Critical Examination of English Space and Time Theories from Henry More to Bishop Berkeley*. Bronxville, NY: Sarah Lawrence College.

Ball, B.W. 2008. *The Soul Sleepers: Christian Mortalism from Wycliffe to Priestley*. Cambridge: James Clarke & Co.

Banchetti-Robino, M.P. 2020. *The Chemical Philosophy of Robert Boyle: Mechanicism, Chymical Atoms, and Emergence*. New York: Oxford University Press.

Barrow, I. 1685. *Lectiones Mathematicæ XXIII*. London.

Barrow, I. 1734. *The Usefulness of Mathematical Learning*. London.

Bate, J. 1634. *The Mysteries of Nature and Art*. London.

Belkind, O. 2020. "Unnatural Acts: The Transition from Natural Principles to Laws of Nature in Early Modern Science," *Studies in History and Philosophy of Science* 81: 62–73.

Benítez, L. 2006. "Newton's Notion of Matter in the 'De aere et aethere'," *Lumen* 25: 17–28.

Benítez, L. 2012. "The New Ontology in the Wake of Newton's and Clarke's Natural Philosophy," in G. Hurtado and O. Nudler (eds.) *The Furniture of the World: Essays in Ontology and Metaphysics*. Leiden: Brill. 111–126.

Benítez, L. and J.A. Robles. 2006. *De Newton y los newtonianos: entre Descartes y Berkeley*. Bernal, Quilmes: Universidad Nacional de Quilmes Editorial.

Bennett, J. 2005. "God and Matter in Locke: An Exposition of *Essay* IV.10," in C. Mercer and E. O'Neill (eds.) *Early Modern Philosophy: Mind, Matter, and Metaphysics*. Oxford: Oxford University Press. 161–182.

Bennett, J. and P. Remnant. 1978. "How Matter Might at First Be Made," *Canadian Journal of Philosophy* 8(sup. 1): 1–11.

Berkeley, G. 1998. *A Treatise Concerning the Principles of Human Knowledge*. J. Dancy (ed.). New York: Oxford University Press.

Bertoloni Meli, D. 2005. "Who is Afraid of Centrifugal Force?" *Early Science and Medicine* 10(4): 535–543.

Biener, Z. 2017. "*De gravitatione* Reconsidered: The Changing Significance of Experimental Evidence for Newton's Metaphysics of Space," *Journal of the History of Philosophy* 55(4): 583–608.

Biener, Z. 2020. "Definitions *more geometrarum* and Newton's Scholium on Space and Time," *Studies in History and Philosophy of Modern Physics* 72: 179–191.

Biener, Z. and E. Schliesser. 2017. "The Certainty, Modality, and Grounding of Newton's Laws," *The Monist* 100(3): 311–325.

Biener, Z. and C. Smeenk. 2012. "Cotes's Queries: Newton's Empiricism and Concepts of Matter," in E. Schliesser and A. Janiak (eds.) *Newton and Empiricism*. Cambridge: Cambridge University Press. 105–137.

Blundeville, T. 1599. *The Art of Logike*. London.

Boeker, R. 2021. *Locke on Persons and Personal Identity*. New York: Oxford University Press.

Boyle, R. 1662. *A Defence of the Doctrine Touching the Spring and Weight of the Air*. London.

Boyle, R. 1664. *Experiments and Considerations Touching Colours*. London.

Boyle, R. 1999–2000. *The Works of Robert Boyle*. M. Hunter and E.B. Davis (eds.). London: Pickering & Chatto.

Brackenridge, J.B. 1993. "The Locke/Newton Manuscripts Revisited: Conjugates, Curvatures, & Conjectures," *Archives internationales d'histoire des sciences* 43: 280–292.

Brading, K. *Forthcoming*. "Newton on Body," in E. Schliesser and C. Smeenk (eds.) *The Oxford Handbook of Newton*. New York: Oxford University Press.

Brading, K. 2025. "In Search of the Historical Newton," *History of European Ideas* 51(3): 639–645.

Brading, K. 2011. "On Composite Systems: Descartes, Newton, and the Law-Constitutive Approach," in D. Jalobeanu and P. Anstey (eds.) *Vanishing Matter and the Laws of Nature*. New York: Routledge. 130–152.

Brading, K. 2012. "Newton's Law-Constitutive Approach to Bodies," in A. Janiak and E. Schliesser (eds.) *Interpreting Newton*. Cambridge: Cambridge University Press. 13–32.

Brading, K. 2013. "Three Principles of Unity in Newton" *Studies in History and Philosophy of Science* 44: 408–415.

Brading, K. 2019. "A Note on Rods and Clocks in Newton's *Principia*" *Studies in History and Philosophy of Science* 67: 160–166.

Brading, K. and M. Stan. 2021. "How Physics Flew the Philosophers' Nest," *Studies in History and Philosophy of Science* 88: 312–320.

Brading, K. and M. Stan. 2023. *Philosophical Mechanics in the Age of Reason*. New York: Oxford University Press.

Brewster, D. 1860. *Memoirs of the Life, Writings, and Discoveries of Sir Isaac Newton*, 2nd edition, 2 volumes. Edinburgh: Edmonston and Douglas.

Buchwald, J.Z. and M. Feingold. 2013. *Newton and the Origin of Civilization*. Princeton: Princeton University Press.

Burgersdijk, F. 1640. *Institutionum Metaphysicarum*. Leiden.

Burns, N.T. 1972. *Christian Mortalism from Tyndale to Milton*. Cambridge, MA: Harvard University Press.

Burthogge, R. 1699. *Of the Soul of the World and of Particular Souls*. London.

Calloway, K. 2014. *Natural Theology in the Scientific Revolution: God's Scientists*. London: Pickering & Chatto.

Carriero, J. 1990. "Newton on Space and Time: Comments on J.E. McGuire," in P. Bricker and R.I.G. Hughes (eds.) *Philosophical Perspectives on Newtonian Science*. Cambridge, MA: MIT Press. 109–133.

Casini, P. 1984. "Newton: The Classical Scholia" *History of Science* 22(1): 1–46.

Chalmers, A. 2009. *The Scientist's Atom and the Philosopher's Stone*. Dordrecht: Springer.

Charleton, W. 1652. *The Darknes of Atheism Dispelled*. London.

Charleton, W. 1654. *Physiologia Epicuro-Gassendo-Charltoniana*. London.

Chen, E.D. 2020. "Newton's Early Metaphysics of Body: Impenetrability, Action at a Distance, and Essential Gravity," *Studies in History and Philosophy of Physics* 72: 192–204.

Cheyne, G. 1705. *Philosophical Principles of Natural Religion*. London.

Clarke, S. 1998. *A Demonstration of the Being and Attributes of God*. E. Vailati (ed.). New York: Cambridge University Press.

Cohen, I.B. 1958. "Versions of Isaac Newton's First Published Paper," *Archives internationales d'histoire des sciences* 11: 357–375.

Cohen, I.B. 1966. "Hypotheses in Newton's Philosophy," *Physis* 8(2): 163–184.

Cohen, I.B. 1969. "Isaac Newton's *Principia*, the Scriptures, and Divine Providence," in S. Morgenbesser, P. Suppes, and M. White (eds.) *Philosophy, Science, and Method: Essays in Honor of Ernst Nagel*. New York: St. Martin's Press. 523–548.
Cohen, I.B. 1990. "Halley's Two Essays on Newton's *Principia*," in N.J.W. Thrower (ed.) *Standing on the Shoulders of Giants*. Berkeley: University of California Press. 91–108.
Colliber, S. 1718. *An Impartial Inquiry into the Existence and Nature of God*. London.
Collins, J.R. 2020. *In the Shadow of Leviathan: John Locke and the Politics of Conscience*. New York: Cambridge University Press.
Conn, C. 1999. "Two Arguments for Lockean Four-Dimensionalism," *British Journal for the History of Philosophy* 7(3): 429–446.
Connolly, P.J. 2014. "Newton and God's Sensorium," *Intellectual History Review* 24(2): 185–201.
Connolly, P.J. 2015a. "Space Before God? A Problem in Newton's Metaphysics," *Philosophy* 90(1): 83–106.
Connolly, P.J. 2015b. "Lockean Superaddition and Lockean Humility," *Studies in History and Philosophy of Science* 51: 53–61.
Connolly, P.J. 2016. "Maclaurin on Occasionalism: A Reply to Ablondi," *Journal of Scottish Philosophy* 14(1): 125–135.
Connolly, P.J. 2017. "Metaphysics in Richard Bentley's Boyle Lectures," *History of Philosophy Quarterly* 34(2): 155–174.
Connolly, P.J. 2019. "Thinking Matter in Locke's Proof of God's Existence," *Oxford Studies in Early Modern Philosophy* 9: 105–130.
Connolly, P.J. 2021. "Causation and Gravitation in George Cheyne's Newtonian Natural Philosophy," *Studies in History and Philosophy of Science* 81: 145–154.
Connolly, P.J. 2024. "Gravity, Occult Qualities, and Newton's Ontology of Powers," in D. Perler and S. Bender (eds.) *Powers and Abilities in Early Modern Philosophy*. New York: Routledge. 146–166.
Connolly, P.J. 2026. "Locke on the Proper Use of Hypotheses in Natural Philosophy," in L. Simonutti and R. Evangelista (eds.) *Locke and the Sciences*. Turnhout: Brepols.
Coplestone, F. 1959. *A History of Philosophy*, volume 5. New York: The Newman Press.
Costello, W.T. 1958. *The Scholastic Curriculum at Early Seventeenth-Century Cambridge*. Cambridge, MA: Harvard University Press.
Cotes, R. 1738. *Hydrostatical and Pneumatical Lectures*. London.
Cunning, D. 2008. "Malebranche and Occasional Causes," *Philosophy Compass* 3(3): 471–490.
Davis, E.B. 1991. "Newton's Rejection of the 'Newtonian World View': The Role of Divine Will in Newton's Natural Philosophy," *Science and Christian Belief* 3: 103–117.
De Angelis, A. 2021. *Galileo Galilei's 'Two New Sciences': For Modern Readers*. Cham: Springer.
De Smet, R. and K. Verelst. 2001. "Newton's Scholium Generale: The Platonic and Stoic Legacy—Philo, Justus Lipsius and the Cambridge Platonists," *History of Science* 39(1): 1–30.
Dempsey, L. 2006. "Written in the Flesh: Isaac Newton on the Mind-Body Relation," *Studies in History and Philosophy of Science* 37(3): 420–442.
Dempsey, L. 2009. "Thinking-Matter Then and Now: The Evolution of Mind-Body Dualism," *History of Philosophy Quarterly* 26(1): 43–61.
Dempsey, L. 2011. "'A Compound Wholly Mortal': Locke and Newton on the Metaphysics of (Personal) Immortality," *British Journal for the History of Philosophy* 19(2): 241–264.
Dempsey, L. 2014. "Newtonian Idealism: Matter, Perception, and the Divine Will," *Southern Journal of Philosophy* 52(1): 86–112.
Dempsey, L. 2017. "John Locke, 'Hobbist': Of Sleeping Souls and Thinking Matter," *Canadian Journal of Philosophy* 47(4): 454–476.
Dempsey, L. and B. Stoyles. 2010. "Comfort in Annihilation: Three Studies in Materialism and Mortality," *Forum Philosophicum* 15(1): 119–140.
Dhondt, F. and S. Ducheyne. 2021. "Theological and Religious Statements in Isaac Newton's Queries/Quaestiones to the Opticks/Optice," *European Journal of Science and Theology* 17(3): 1–10.
Digby, K. 1644. *Two Treatises*. Paris.

DiSalle, R. 2013. "The Transcendental Method from Newton to Kant," *Studies in History and Philosophy of Science* 44(3): 448–456.
DiSalle, R. 2016. "Newton's Philosophical Analysis of Space and Time" in R. Iliffe and G.E. Smith (eds.) *The Cambridge Companion to Newton*, 2nd edition. New York: Cambridge University Press. 34–60.
Dobbs, B.J.T. 1991. *The Janus Faces of Genius: The Role of Alchemy in Newton's Thought.* Cambridge: Cambridge University Press.
Dobbs, B.J.T. 1993. "'The Unity of Truth': An Integrated View of Newton's Work," in P. Theerman and A.F. Seeff (eds.) *Action and Reaction: Proceedings of a Symposium to Commemorate the Tercentenary of Newton's Principia.* Newark: University of Delaware Press. 105–122.
Domski, M. 2010. "Newton's Empiricism and Metaphysics," *Philosophy Compass* 5(7): 525–534.
Domski, M. 2021. *Newton's Third Rule and the Experimental Argument for Universal Gravity.* New York: Routledge.
Downing, L. 2011. Review of A. Janiak, *Newton as Philosopher, Philosophical Review* 120(1): 124–129.
Downing, L. 2012. "Maupertuis on Attraction as an Inherent Property of Matter," in A. Janiak and E. Schliesser (eds.) *Interpreting Newton.* Cambridge: Cambridge University Press. 280–298.
Downing, L. 2014. "Locke's Metaphysics and Newtonian Metaphysics," in Z. Biener and E. Schliesser (eds.) *Newton and Empiricism.* Oxford: Oxford University Press. 97–118.
Dry, S. 2014. *The Newton Papers: The Strange and True Odyssey of Isaac Newton's Manuscripts.* New York: Oxford University Press.
Ducheyne, S. 2009a. "Understanding (in) Newton's Argument for Universal Gravitation," *Journal for General Philosophy of Science* 40(2): 227–258.
Ducheyne, S. 2009b. "The Flow of Influence: From Newton to Locke . . . And Back," *Rivisita di Storia della Filosofia* 64(2): 245–268.
Ducheyne, S. 2012. *The Main Business of Natural Philosophy: Isaac Newton's Natural-Philosophical Methodology.* Dordrecht: Springer.
Ducheyne, S. 2021. "Towards a Comparative Perspective on Newton's Working Methods," *Early Science and Medicine* 26(2): 152–161.
Duncan, S. 2022. *Materialism from Hobbes to Locke.* New York: Oxford University Press.
Edwards, J. 1699. *The Eternal and Instrinsick Reasons of Good and Evil.* Cambridge.
Elliott, R.W.V. 1957. "Isaac Newton's 'Of a Universal Language,'" *The Modern Language Review* 52(1): 1–18.
Faulconer, J.E. 1995. "Newton, Science, and Causation," *The Journal of Mind and Behavior* 16(1): 77–86.
Feingold, M. 2019. "The Religion of the Young Isaac Newton," *Annals of Science* 76(2): 210–218.
Feingold, M. 2021. "How Important was Religion to Newton's 'Secular' Studies?" *Early Science and Medicine* 26(2): 181–190.
Figala, K. 2002. "Newton's Alchemy," in I.B. Cohen and G. Smith (eds.) *The Cambridge Companion to Newton.* Cambridge: Cambridge University Press. 370–386.
Force, J.E. 1990a. "Newton's God of Dominion: The Unity of Newton's Theological, Scientific, and Political Thought," in J.E. Force and R.H. Popkin (eds.) *Newton and Religion: Context, Nature, and Influence.* Dordrecht: Kluwer. 75–102.
Force, J.E. 1990b. "Newton's 'Sleeping Argument' and the Newtonian Synthesis of Science and Religion," in N.J.W. Thrower (ed.) *Standing on the Shoulders of Giants: A Longer View of Newton and Halley.* Berkeley and Los Angeles: University of California Press. 109–127.
Force, J.E. 1994. "The God of Abraham and Isaac (Newton)," in J.E. Force and R.H. Popkin (eds.) *The Books of Nature and Scripture: Recent Essays on Natural Philosophy, Theology and Biblical Criticism in the Netherlands of Spinoza's Time and the British Isles of Newton's Time.* Dordrecht: Kluwer. 179–200.
Force, J.E. 1999. "'Children of the Resurrection' and 'Children of the Dust': Confronting Mortality and Immortality with Newton and Hume," in J.E. Force (ed.) *Everything Connects: In Conference with Richard H. Popkin: Essays in His Honor.* Leiden: Brill. 119–142.

Force, J.E. 2000. "The Nature of Newton's 'Holy Alliance' between Science and Religion: From the Scientific Revolution to Newton (and Back Again)," in M.J. Osler (ed.) *Rethinking the Scientific Revolution*. New York: Cambridge University Press. 247–270.
Freddoso, A.J. 1988. "Medieval Aristotelianism and the Case Against Secondary Causation in Nature," in T.V. Morris (ed.) *Divine and Human Action: Essays in the Metaphysics of Theism*. Ithaca: Cornell University Press. 74–118.
Freddoso, A.J. 1991. "God's General Concurrence with Secondary Causes: Why Conservation Is not Enough," *Philosophical Perspectives* 5: 553–585.
Freddoso, A.J. 1994. "God's General Concurrence with Secondary Causes: Pitfalls and Prospects," *American Catholic Philosophical Quarterly* 68(2): 131–156.
Frost, G. 2014. "Peter Olivi's Rejection of God's Concurrence," *British Journal for the History of Philosophy* 22(4): 655–679.
Gabbey, A. 2001. "Disciplinary Transformations in the Age of Newton: The Case of Metaphysics," in W. Lefèvre (ed.) *Between Leibniz, Newton, and Kant*. Dordrecht: Kluwer. 3–23.
Gabbey, A. 2011. "The Term *Materia* in Newton and in the Newtonian Tradition," in D. Giovannozzi and M. Veneziani (eds.) *Materia: XIII Colloquio Internazionale*. Florence: Leo S. Olshcki Editore. 423–445.
Gabbey, A. 2016. "Newton, Active Powers, and the Mechanical Philosophy," in R. Iliffe and G.E. Smith (eds.) *The Cambridge Companion to Newton*. 2nd edition. Cambridge: Cambridge University Press. 421–453.
Gal, O. 2005. "The Invention of Celestial Mechanics," *Early Science and Medicine* 10(4): 529–534.
Garber, D. 1983. "Mind, Body, and the Laws of Nature in Descartes and Leibniz," *Midwest Studies in Philosophy* 8: 103–133.
Gascoigne, J. 1989. *Cambridge in the Age of the Enlightenment*. Cambridge: Cambridge University Press.
Gascoigne, J. 2004. "Greene, Robert (c. 1678–1730), natural philosopher," *Oxford Dictionary of National Biography*.
Gillespie, G. 1637. *A Dispute Against the English-Popish Ceremonies*. Leiden.
Glanvill, J. 1668. *Plus Ultra*. London.
Gorham, G. 2011a. "Newton on God's Relation to Space and Time: The Cartesian Framework," *Archiv für Geschichte der Philosohpie* 93(3): 281–320.
Gorham, G. 2011b. "How Newton Solved the Mind-Body Problem," *History of Philosophy Quarterly* 28(1): 21–44.
Gorham, G. 2012. "'The Twin-Brother of Space': Spatial Analogy in the Emergence of Absolute Time," *Intellectual History Review* 22(1): 23–39.
Grant, E. 1981. *Much Ado About Nothing: Theories of Space and Vacuum from the Middle Ages to the Scientific Revolution*. New York: Cambridge University Press.
Greene, R. 1727. *The Principles of the Philosophy of Expansive and Contractive Forces*. Cambridge.
Greenham, P. 2017. "Clarifying Divine Discourse in Early Modern Science: Divinity, Physico-Theology, and Divine Metaphysics in Newton's Chymistry," *The Seventeenth Century* 32(2): 191–215.
Guerlac, H. 1983. "Theological Voluntarism and Biological Analogies in Newton's Physical Thought," *Journal of the History of Ideas* 44(2): 219–229.
Gueroult, M. 1984. *Descartes' Philosophy Interpreted According to the Order of Reasons*, 2 volumes. R. Ariew (trans.). Minneapolis: University of Minnesota Press.
Guicciardini, N. 2005. "Reconsidering the Hooke-Newton Debate on Gravitation: Recent Results," *Early Science and Medicine* 10(4): 510–517.
Guicciardini, N. 2013. "The Role of Musical Analogies in Newton's Optical and Cosmological Work," *Journal of the History of Ideas* 74(1): 45–67.
Guicciardini, N. 2018. *Isaac Newton and Natural Philosophy*. London: Reaktion Books.
Guicciardini, N. 2021. "Lessons for the Historian of Newtonian Mathematics," *Early Science and Medicine* 26(2): 124–135.
Guzzardi, L. 2020. *Ruggiero Boscovich's Theory of Natural Philosophy*. Cham: Springer.
Hall, A.R. 1998. "Isaac Newton and the Aerial Nitre," *Notes and Records of the Royal Society of London* 52(1): 51–61.

Hall, A.R. and M.B. Hall. 1962. *Unpublished Scientific Papers of Sir Isaac Newton*. Cambridge: Cambridge University Press.

Hamou, P. 2014. "Vision, Color, and Method in Newton's *Opticks*," in Z. Biener and E. Schliesser (eds.) *Newton and Empiricism*. New York: Oxford University Press. 66–93.

Harrison, P. 1995. "Newtonian Science, Miracles, and the Laws of Nature," *Journal of the History of Ideas* 56(4): 531–553.

Harrison, J. and P. Laslett. 1971. *The Library of John Locke*, 2nd edition. Oxford: Clarendon Press.

de Hatzfeld, J.C.F. 1724. *The Case of the Learned Represented*. London.

Hazelwood, C. 2023. "Newton's 'Law-First' Epistemology and 'Matter-First' Metaphysics," *Studies in History and Philosophy of Science* 101: 40–47.

Heimann, P.M. 1973a. "Newtonian Natural Philosophy and the Scientific Revolution," *History of Science* 11(1): 1–7.

Heimann, P.M. 1973b. "'Nature is a Perpetual Worker': Newton's Aether and Eighteenth-Century Natural Philosophy," *Ambix* 20(1): 1–25.

Heimann, P.M. 1978. "Voluntarism and Immanence: Conceptions of Nature in Eighteenth-Century Thought," *Journal of the History of Ideas* 39(2): 271–283.

Heimann, P.M. and J.E. McGuire. 1971. "Newtonian Forces and Lockean Powers: Concepts of Matter in Eighteenth-Century Thought," *Historical Studies in the Physical Sciences* 3: 233–306.

Henry, J. 1986. "Occult Qualities and the Experimental Philosophy: Active Principles in Pre-Newtonian Matter Theory," *History of Science* 24(4): 335–381.

Henry, J. 1988. "Newton, Matter, and Magic," in J. Fauvel, R. Flood, M. Shortland, and R. Wilson (eds.) *Let Newton Be!* New York: Oxford University Press. 127–145.

Henry, J. 2011. "Gravity and *De gravitatione*: The Development of Newton's Ideas on Action at a Distance," *Studies in History and Philosophy of Science* 42(1): 11–27.

Henry, J. 2020a. "Primary and Secondary Causation in Samuel Clarke's and Isaac Newton's Theories of Gravity," *Isis* 111(3): 542–561.

Henry, J. 2020b. "Newton, the Sensorium of God, and the Cause of Gravity," *Science in Context* 33(3): 329–251.

Henry, J. 2021. "Reassessing the Wider Aspects of Newton's Thought – A Symposium," *Early Science and Medicine* 26(2): 117–123.

Henry, J. 2023. "Newton's 'De Aere et Aethere' and the Introduction of Interparticulate Forces into his Physics," *Annals of Science* 80(3): 232–267.

Henry, J. 2024a. Review of Schliesser, E. *Newton's Metaphysics: Essays*, *HOPOS* 14(1): 246–250.

Henry, J. 2024b. "'Like Nets or Cobwebs': Kenelm Digby, Isaac Newton, and the Problem of Rarefaction," *British Journal for the History of Science* 57(3): 447–466.

Henry, J. and J.E. McGuire. 2018. "Voluntarism and Panentheism: The Sensorium of God and Isaac Newton's Theology," *The Seventeenth Century* 33(5): 587–612.

Hill, B. 2003. "Newton's *De Gravitatione et Aequipondio Fluidorum* and Lockean Four-Dimensionalism," *British Journal for the History of Philosophy* 11(2): 309–321.

Holden, T. 2004. *The Architecture of Matter: Galileo to Kant*. New York: Oxford University Press.

Hooke, R. 1665. *Micrographia*. London.

Hooke, R. 1674. *An Attempt to Prove the Motion of the Earth by Observation*. London.

Hooke, R. 1678. *Lectures de Potentia Restitutiva*. London.

Hoskin, M. 1988. "Newton and the Beginnings of Stellar Astronomy," in G.V. Coyne S.J., H. Heller, and J. Życiński (eds.) *Newton and the New Direction in Science*. Vatican City: Specola Vaticana. 55–63.

Hurlbutt, R.H. 1965. *Hume, Newton, and the Design Argument*. Lincoln: University of Nebraska Press.

Hutton, S. 1999. "The Seven Trumpets and the Seven Vials: Apocalypticism and Christology in Newton's Theological Writings," in J.E. Force and R.H. Popkin (eds.) *Newton and Religion: Context, Nature, and Influence*. Dordrecht: Kluwer. 165–178.

Iliffe, R. 1995. "'That puzleing Problem': Isaac Newton on the Political Physiology of the Self," *Medical History* 39(4): 433–458.

Iliffe, R. 1999. "Those 'Whose Business it is to Cavill': Newton's Anti-Catholicism," in J.E. Force and R.H. Popkin (eds.) *Newton and Religion: Context, Nature, and Influence*. Dordrecht: Kluwer. 97–119.

Iliffe, R. 2004. "Abstract Considerations: Disciplines and the Incoherence of Newton's Natural Philosophy," *Studies in History and Philosophy of Science* 35(3): 427–454.

Iliffe, R. 2006. *Early Biographies of Isaac Newton, 1660-1885*, 2 volumes. London: Pickering & Chatto.

Iliffe, R. 2007. *Newton: A Very Short Introduction*. New York: Oxford University Press.

Iliffe, R. 2017. *Priest of Nature: The Religious Worlds of Isaac Newton*. Oxford: Oxford University Press.

Iliffe, R. 2021. "Response to Comments on *Priest of Nature*," *Early Science and Medicine* 26(2): 191–199.

Jackson, T. 1654. *An Exact Collection of the Works of Doctor Jackson*. London.

Jacovides, M. 2017. *Locke's Image of the World*. New York: Oxford University Press.

Jalobeanu, D. 2007. "Space, Bodies and Geometry: Some Sources of Newton's Metaphysics," *Zeitsprünge, Forschungen zur Frühen Neuzeit* 11: 81–113.

Jalobeanu, D. 2011. "The Cartesians of the Royal Society: The Debate Over Collisions and the Nature of Body (1668–1670)," in D. Jalobeanu and P.R. Anstey (eds.) *Vanishing Matter and the Laws of Motion*. New York: Routledge. 103–129.

Janiak, A. 2000. "Space, Atoms, and Mathematical Divisibility in Newton," *Studies in History and Philosophy of Science* 31(2): 203–230.

Janiak, A. 2007. "Newton and the Reality of Force," *Journal of the History of Philosophy* 45(1): 127–147.

Janiak, A. 2008. *Newton as Philosopher*. New York: Cambridge University Press.

Janiak, A. 2010. "Substance and Action in Descartes and Newton," *The Monist* 93(4): 657–677.

Janiak, A. 2012. "Newton and Descartes: Theology and Natural Philosophy," *Southern Journal of Philosophy* 50(3): 414–435.

Janiak, A. 2013. "Three Concepts of Causation in Newton," *Studies in History and Philosophy of Science* 44(3): 396–407.

Janiak, A. 2015. "Space and Motion in Nature and Scripture: Galileo, Descartes, and Newton," *Studies in History and Philosophy of Science* 51: 89–99.

Janiak, A. 2023. "A Tale of Two Forces: Metaphysics and its Avoidance in Newton's *Principia*," in M. Stan and C. Smeenk (eds.) *Theory, Evidence, Data: Themes from George E. Smith*. Cham: Springer. 223–242.

Joalland, M. and S. Mandelbrote. 2016. "Isaac Newton Learns Hebrew: Samuel Johnson's *Nova cui hebræi tabella*," *Notes and Records* 70(1): 9–21.

Jolley, N. 2015. *Locke's Touchy Subjects: Materialism and Immortality*. New York: Oxford University Press.

Joy, L.S. 2012. "Dispositional Explanations: Boyle's Problem, Newton's Solution, Hume's Response," in A. Janiak and E. Schliesser (eds.) *Interpreting Newton*. New York: Cambridge University Press. 320–341.

Kargon, R. 1966. *Atomism in England from Harriot to Newton*. Oxford: Clarendon Press.

Kassler, J.C. 2018. *Newton's Sensorium: Anatomy of a Concept*. Cham: Springer.

Keill, J. 1699. *An Examination of the Reflections on the Theory of the Earth*. Oxford.

Kemeny, M. 2017. "What Motion *is*: William Neile and the Laws of Motion," *Annals of Science* 74(3): 179–191.

Kerszberg, P. 1986. "The Cosmological Question in Newton's Science," *Osiris* 2: 69–106.

Khamara, E.J. 2006. *Space, Time, and Theology in the Leibniz-Newton Controversy*. Frankfurt: Ontos.

Kochiras, H. 2009. "Gravity and Newton's Substance Counting Problem," *Studies in History and Philosophy of Science* 40: 267–280.

Kochiras, H. 2011. "Gravity's Cause and Substance Counting: Contextualizing the Problems," *Studies in History and Philosophy of Science* 42: 167–184.

Kochiras, H. 2012. "Spiritual Presence and Dimensional Space Beyond the Cosmos," *Intellectual History Review* 22(1): 41–68.

Kochiras, H. 2013. "By ye Divine Arm: God and Substance in *De gravitatione*," *Religious Studies* 49(3): 327–356.

Kollerstrom, N. 1999. "The Path of Halley's Comet, and Newton's Late Apprehension of the Law of Gravity," *Annals of Science* 56(4): 331–356.

Koyré, A. 1965. *Newtonian Studies*. Chicago: University of Chicago Press.

Koyré, A. and I.B. Cohen. 1961. "The Case of the Missing *Tanquam*: Leibniz, Newton & Clarke," *Isis* 52(4): 555–566.

Koyré, A. and I.B. Cohen. 1962. "Newton and the Leibniz-Clarke Correspondence," *Archives internationales d'histoire des sciences* 15(1): 63–126.

Kulstad, M. 2008. "Newton, Spinoza, Stoics, and Others: A Battle Line in Leibniz's Wars of (Natural) Religion," *Leibniz Review* 18: 81–121.

Leibniz, G.W. 1875–1890. *Die Philosophichen Schriften*, 7 volumes. Berlin: Weidmannsche Buchhandlung.

Leibniz, G.W. 2008. *The Art of Controversies*. M. Dascal (ed. and trans.). Dordrecht: Springer.

Leibniz, G.W. and S. Clarke. 2000. *Correspondence*. R. Ariew (ed. and trans.). Indianapolis: Hackett Publishing Company.

Leshem, A. 2003. *Newton on Mathematics and Spiritual Purity*. Dordrecht: Kluwer.

Levitin, D. 2016. "Newton and Scholastic Philosophy," *British Journal for the History of Science* 49(1): 53–77.

Levitin, D. 2021. "Isaac Newton's '*De gravitatione et aequipondio fluidorum*': Its Purpose in Historical Context," *Annals of Science* 78(2): 133–161.

Levitin, D. 2022. *The Kingdom of Darkness: Bayle, Newton, and the Emancipation of the European Mind from Philosophy*. New York: Cambridge University Press.

Levitin, D. 2025. "Reply to Critics: Studying Early Modern Philosophers as if They Were Human Beings" *History of European Ideas* 51(3): 646–657.

Levitin, D. and S. Mandelbrote. 2019. "Becoming Heterodox in 17th-Century Cambridge: The Case of Isaac Newton," in N. Hardy and D. Levitin (eds.) *Confessionalisation and Erudition in Early Modern Europe: An Episode in the History of the Humanities*. Oxford: Oxford University Press. 301–394.

Line, F. 1661. *Tractatus de Corporum Inseparabilitate*. London.

Locke, J. 1690. *An Essay concerning Humane Understanding*. London.

Locke, J. 1729. *Essai philosophique concernant l'entendement humain*. P. Coste (trans.). Amsterdam.

Locke, J. 1735. *Essai philosophique concernant l'entendement humain*. P. Coste (trans.). Amsterdam.

Locke, J. 1823. *The Works of John Locke*, 10 volumes. London.

Locke, J. 1975. *An Essay concerning Human Understanding*. P.H. Nidditch (ed.). Oxford: Clarendon Press.

Locke, J. 1976–1989. *The Correspondence of John Locke*, 8 volumes. E.S. de Beer (ed.). Oxford: Clarendon Press.

Locke, J. 1997. *Political Essays*. M. Goldie (ed.). New York: Cambridge University Press.

Lucci, D. 2021. *John Locke's Christianity*. New York: Cambridge University Press.

Lüthy, C. 2024. "Giving up Philosophy?" *History of European Ideas* 51(3): 628–631.

MacLaurin, C. 1748. *An Account of Sir Isaac Newton's Philosophical Discoveries*. London.

Maddison, R.E.W. 1969. *The Life of the Honourable Robert Boyle, F.R.S.* London: Taylor & Francis Ltd.

Magirus, J. 1642. *Physiologiæ Peripateticæ*. Cambridge.

Maimonides, M. 1963. *The Guide of the Perplexed*, 2 volumes. S. Pines (trans.). Chicago: The University of Chicago Press.

Malcolm, N. 2002. "Hobbes and Roberval," in N. Malcolm (ed.) *Aspects of Hobbes*. New York: Oxford University Press. 156–199.

Mamiani, M. 1991. "The Rhetoric of Certainty: Newton's Method in Science and in the Interpretation of the Apocalypse," in M. Pera and W.R. Shea (eds.) *Persuading Science: The Art of Scientific Rhetoric*. Canton, MA: Science History Publications. 157–172.

Mamiani, M. 2000. "To Twist the Meaning: Newton's *Regulae Philosophandi* Revisited," in J.Z. Buchwald and I.B. Cohen (eds.) *Isaac Newton's Natural Philosophy*. Cambridge, MA: MIT Press. 3–14.

Mamiani, M. 2002. "Newton on Prophecy and the Apocalypse," in I.B. Cohen and G.E. Smith (eds.) *The Cambridge Companion to Newton*. New York: Cambridge University Press. 387–408.

Mandelbrote, S. 1994. "Isaac Newton and Thomas Burnet: Biblical Criticism and the Crisis of Late Seventeenth-Century England," in J.E. Force and R.H. Popkin (eds.) *The Books of Nature and Scripture: Recent Essays on Natural Philosophy, Theology and Biblical Criticism in the Netherlands of Spinoza's Time and the British Isles of Newton's Time*. Dordrecht: Kluwer. 149–178.

Mandelbrote, S. 2006a. "'Than This Nothing Can Be Plainer': Isaac Newton Reads the Fathers," in G. Frank, T. Leinkauf, and M. Wriedt (eds.) *Die Patristik in der Frühen Neuzeit*. Stuttgart: Frommann-Holzboog. 277–298.

Mandelbrote, S. 2006b. "Isaac Newton and the Flood," in M. Muslow and J. Assman (eds.) *Sintflut und Gedächtnis: Erinnern und Vergessen des ursprungs*. Munich: Wilhelm Fink Verlag. 337–353.

Mandelbrote, S. 2021. "Beyond a Boundary: Reflections on Newton the Historian, Theologian, and Alchemist," *Early Science and Medicine* 26(2): 136–151.

Manuel, F.E. 1963. *Isaac Newton: Historian*. Cambridge, MA: Belknap Press of Harvard University Press.

Manuel, F.E. 1974. *The Religion of Isaac Netwon*. Oxford: Clarendon Press.

Marmodoro, A. 2015. "Gregory of Nyssa on the Creation of the World," in B.D. Prince and A. Marmodoro (eds.) *Causation and Creation in Late Antiquity*. Cambridge: Cambridge University Press. 94–110.

McCann, E. 2001. "Locke's Theory of Substance Under Attack!" *Philosophical Studies* 106(1): 87–105.

McGuire, J.E. 1966. "Body and Void in Newton's *De Mundi Systemate*: Some New Sources," *Archive for History of Exact Sciences* 3(3): 206–248.

McGuire, J.E. 1968a. "Force, Active Principles, and Newton's Invisible Realm," *Ambix* 15(3): 154–208.

McGuire, J.E. 1968b. "The Origin of Newton's Doctrine of Essential Qualities," *Centaurus* 12(4): 233–260.

McGuire, J.E. 1970. "Newton's 'Principles of Philosophy': An Intended Preface for the 1704 'Opticks' and a Related Draft Fragment," *British Journal for the History of Science* 5(2): 178–186.

McGuire, J.E. 1978a. "Existence, Actuality and Necessity: Newton on Space and Time," *Annals of Science* 35(5): 463–508.

McGuire, J.E. 1978b. "Newton on Place, Time, and God: An Unpublished Source," *British Journal for the History of Science* 11(2): 114–129.

McGuire, J.E. 1982. "Space, Infinity and Indivisibility: Newton on the Creation of Matter," in Z. Bechler (ed.) *Contemporary Newtonian Research*. Dordrecht: D. Reidel 145–190.

McGuire, J.E. 1983. "Space, Geometrical Objects, and Infinity: Newton and Descartes on Extension," in W.R. Shea (ed.) *Nature Mathematicized*. Dordrecht: Reidel. 69–112.

McGuire, J.E. 2000. "The Fate of the Date: The Theology of Newton's *Principia* Revisited," in M.J. Osler (ed.) *Rethinking the Scientific Revolution*. New York: Cambridge. 271–295.

McGuire, J.E. 2007. "A Dialogue with Descartes: Newton's Ontology of True and Immutable Natures," *Journal of the History of Philosophy* 45(1): 103–125.

McGuire, J.E. 2017. "Natural Motion and Its Causes: Newton on the 'Vis Insita' of Bodies," in J.G. Lennox and M.L. Gill (eds.) *Self-Motion: From Aristotle to Newton*. Princeton: Princeton University Press. 305–329.

McGuire, J.E. and P.M. Rattansi. 1966. "Newton and the 'Pipes of Pan'," *Notes and Records of the Royal Society of London* 21(2): 108–143.

McGuire, J.E. and E. Slowik. 2012. "Newton's Ontology of Omnipresence and Infinite Space," *Oxford Studies in Early Modern Philosophy* 6: 279–308.

McGuire, J.E. and M. Tamny. 1985. "Newton's Astronomical Apprenticeship: Notes of 1664/5," *Isis* 76(3): 349–365.

McMullin, E. 1978. *Newton on Matter and Activity*. Notre Dame, IN: Notre Dame University Press.
McMullin, E. 1985. "The Significance of Newton's *Principia* for Empiricism," in M.J. Osler and P.L. Farber (eds.) *Religion, Science and Worldview: Essays in Honor of Richard S. Westfall*. Cambridge: Cambridge University Press. 167–187.
Melamed, Y.Y. *Manuscript*. "A Concise Grammar of Pantheism."
Melamed, Y.Y. 2018. "Cohen, Spinoza, and the Nature of Pantheism," *Jewish Studies Quarterly* 25(2): 171–180.
Miller, D.M. 2009. "Qualities, Properties, and Laws in Newton's Induction," *Philosophy of Science* 76(5): 1052–1063.
Milton, J. 2008. "Pierre Coste, John Locke, and the Third Earl of Shaftesbury," in S. Hutton and P. Schuurman (eds.) *Studies on Locke: Sources, Contemporaries, and Legacy*. Dordrecht: Springer. 195–223.
Mintz, S. 1962. *The Hunting of Leviathan: Seventeenth-Century Reactions to the Materialism and Moral Philosophy of Thomas Hobbes*. Cambridge: Cambridge University Press.
More, H. 1650. *Observations Upon Anthroposophia Theomagica*. London
More, H. 1653. *An Antidote Against Atheisme*. London.
More, H. 1659. *The Immortality of the Soul*. London.
More, H. 1668. *Divine Dialogues*. London.
More, H. 1995. *Manual of Metaphysics*. A. Jacob (ed. and trans.). New York: Olms.
Murray, G., W. Harper, and C. Wilson. 2011. "Huygens, Wren, Wallis, and Newton on Rules of Impact and Reflection," in D. Jalobeanu and P.R. Anstey (eds.) *Vanishing Matter and the Laws of Motion*. New York: Routledge. 153–191.
Nadler, S. 2000. "Malebranche on Causation," in S. Nadler (ed.) *The Cambridge Companion to Malebranche*. Cambridge: Cambridge University Press. 112–138.
Nauenberg, M. 2005. "Hooke's and Newton's Contributions to the Early Development of Orbital Dynamics and the Theory of Universal Gravitation," *Early Science and Medicine* 10(4): 518–528.
Nerlich, G. 2005. "Can Parts of Space Move? On Paragraph Six of Newton's Scholium," *Erkenntnis* 62(1): 119–135.
Newlands, S. 2024. "Regis's Sweeping and Costly Anti-Spinozism," *Journal of the History of Philosophy* 62(2): 211–238.
Newman, W.R. 2002. "The Background to Newton's Chymistry," in I.B. Cohen and G. Smith (eds.) *The Cambridge Companion to Newton*. Cambridge: Cambridge University Press. 358–369.
Newman, W.R. 2016. "A Preliminary Reassessment of Newton's Alchemy," in R. Iliffe and G.E. Smith (eds.) *The Cambridge Companion to Newton*, 2nd edition. Cambridge: Cambridge University Press. 454–484.
Newman, W.R. 2019. *Newton the Alchemist: Science, Enigma, and the Quest for Nature's "Secret Fire."* Princeton: Princeton University Press.
Newman, W.R. 2021. "A Reply to the Four Reviewers," *Early Science and Medicine* 26(2): 200–205.
Newton, I. 1728. *De Mundi Systemate*. London.
Newton, I. 1731. *A Treatise of the System of the World*. London.
Newton, I. 1985. *De la Gravitation ou les Fondements de la Mécanique Classique*. M.-F. Biarnais (ed. and trans.). Paris: Les Belles Lettres.
Newton, I. 1988. *Über die Gravitation*. G. Böhme (ed. and trans.). Frankfurt: Klostermann.
Nolan, L. 1997. "Reductionism and Nominalism in Descartes's Theory of Attributes," *Topoi* 16(2): 129–140.
Normore, C. 2007. "Descartes and the Metaphysics of Extension," in J. Broughton and J. Carriero (eds.) *A Companion to Descartes*. Malden, MA: Blackwell. 271–287.
Nuovo, V. 2017. *John Locke: The Philosopher as Christian Virtuoso*. Oxford: Clarendon Press.
Ochoa, F. 2005. "Newton y el Dios del Dominio: Teología Voluntarista ilustrada en los conceptos Espacio Absoluto, Tiempo Absoluto y Gravitación Universal," *Estudios de Filosofía* 31: 105–126.
Palmerino, C.R. 2011. "The Isomorphism of Space, Time and Matter in Seventeenth-Century Natural Philosophy," *Early Science and Medicine* 16(4): 296–330.

Palmerino, C.R. 2013. "The Composition of Space, Time and Matter According to Isaac Newton and John Keill," in D. Garber and S. Roux (eds.) *The Mechanization of Natural Philosophy*. Dordrecht: Springer. 117–142.

Palter, R. 1987. "Saving Newton's Text: Documents, Readers, and the Ways of the World," *Studies in History and Philosophy of Science* 18(4): 385–439.

Parker, S. 1666. *A Free and Impartial Censure of the Platonick Philosophie*. Oxford.

Pasnau, R. 2011. *Metaphysical Themes, 1274–1671*. New York: Oxford University Press.

Perler, D. 2024. "Suárez on Powers and Abilities as Inner Causes," in D. Perler and S. Bender (eds.) *Powers and Abilities in Early Modern Philosophy*. New York: Routledge. 23–42.

Peterman, A. *Forthcoming*. "Newton and Spinoza" in E. Schliesser and C. Smeenk (eds.) *The Oxford Handbook of Newton*. New York: Oxford University Press.

Peterman, A. 2021. "The World Soul in Early Modern Philosophy," in J. Wilberding (ed.) *World Soul: A History*. New York: Oxford University Press. 186–222.

Popkin, R.H. 1988. "Newton's Biblical Theology and his Theological Physics," in P.B. Cheurer and G. Debrock (eds.) *Newton's Scientific and Philosophical Legacy*. Dordrecht: Kluwer. 81–97.

Popkin, R.H. 1992. "Newton and Maimonides," in R.H. Popkin (ed.) *The Third Force in Seventeenth-Century Thought*. Leiden: Brill. 189–202.

Power, H. 1664. *Experimental Philosophy*. London.

Price, D.J.deS. 1960. "Newton in a Church Tower: The Discovery of an Unknown Book of Isaac Newton," *The Yale University Library Gazette* 34(3): 124–126.

Rattansi, P. 1988. "Newton and the Wisdom of the Ancients," in J. Fauvel, R. Flood, M. Shortland, and R. Wilson (eds.) *Let Newton Be!* Oxford: Oxford University Press. 185–201.

Reid, J. 2007. "The Evolution of Henry More's Theory of Divine Absolute Space," *Journal of the History of Philosophy* 45(1): 79–102.

Reid, J. 2012. *The Metaphysics of Henry More*. Dordrecht: Springer.

Reid, J. 2022. "Newton and the Cambridge Platonists," in D. Jalobeanu and C.T. Wolfe (eds.) *Encyclopedia of Early Modern Philosophy and the Sciences*. Cham: Springer. 1493–1504.

Reilly, P.C. *Francis Line S.J.: An Exiled English Scientist, 1595–1675*. Rome: Institutum Historicum.

Rickless, S. 2020. "Locke on the Probability of the Mind's Immateriality," *Locke Studies* 20: 1–28.

Robles, J.A. 2000. "Isaac Newton, ¿berkeleyano *avant la lettre*?" *Revista Latinoamericana de Filosofía* 26(2): 199–224.

Robles, J.A. 2001. "Las tesis panteístas de Isaac Newton," *Diánoia: Annuario de filosofía* 46: 3–40.

Robles, J.A. 2006. "Newton y Berkeley, ¿escépticos *malgré lui*?" in L. Benítez and J.A. Robles (eds.) *De Newton y los newtonianos: entre Descartes y Berkeley*. Bernal, Quilmes: Universidad Nacional de Quilmes Editorial. 267–294.

Rogers, G.A.J. 1979. "Locke, Newton, and the Cambridge Platonists on Innate Ideas," *Journal of the History of Ideas* 40(2): 191–205.

Rogers, G.A.J. 1982. "The System of Locke and Newton," in Z. Bechler (ed.) *Contemporary Newtonian Research*. Dordrecht: D. Reidel. 215–238.

Rogers, G.A.J. 1999. "Newton and the Guaranteeing God," in J.E. Force and R.H. Popkin (eds.) *Newton and Religion: Context, Nature, and Influence*. Dordrecht: Kluwer. 221–235.

Rossiter, E. 2016. "Hedonism and Natural Law in Locke's Moral Philosophy," *Journal of the History of Philosophy* 54(2): 203–225.

Russell, B. 1945. *A History of Western Philosophy*. New York: Simon & Schuster.

Sangiacomo, A. 2018. "Samuel Clarke on Agent Causation, Voluntarism, and Occasionalism," *Science in Context* 31(4): 421–456.

Schechtman, A. 2016. "Substance and Independence in Descartes," *Philosophical Review* 125(2): 155–204.

Schliesser, E. 2011a. "Newton's Substance Monism, Distant Action, and the Nature of Newton's Empiricism: Discussion of H. Kochiras' 'Gravity and Newton's Substance Counting Problem,'" *Studies in History and Philosophy of Science* 42: 160–166.

Schliesser, E. 2011b. "Without God: Gravity as a Relational Quality of Matter in Newton's *Treatise*," in D. Jalobeanu and P.R. Anstey (eds.) *Vanishing Matter and the Laws of Motion*. New York: Routledge. 80–100.

Schliesser, E. 2012. "Newton and Spinoza: On Motion and Matter (and God, of Course)" *Southern Journal of Philosophy* 50(3): 436–458.
Schliesser, E. 2013a. "On Reading Newton as an Epicurean: Kant, Spinozism and the Changes to the *Principia*," *Studies in History and Philosophy of Science* 44: 416–428.
Schliesser, E. 2013b. "Newtonian Emanation, Spinozism, Measurement and the Baconian Origin of the Laws of Nature," *Foundations of Science* 18: 449–466.
Schliesser, E. 2013c. "Newton's Philosophy of Time," in H. Dyke and A. Bardon (eds.) *A Companion to the Philosophy of Time*. London: Blackwell. 87–101.
Schliesser, E. 2021. *Newton's Metaphysics: Essays*. New York: Oxford University Press.
Schofield, R.E. 1970. *Mechanism and Materialism: British Natural Philosophy in An Age of Reason*. Princeton: Princeton University Press.
Schüller, V. 2001. "Newton's *Scholia* from David Gregory's Estate on the Propositions IV through IX Book III of his *Principia*," in W. Lefèvre (ed.) *Between Leibniz, Newton, and Kant*. Dordrecht: Kluwer. 213–265.
Seeskin, K. 2017. "Maimonides," in E.N. Zalta (ed.) *Stanford Encyclopedia of Philosophy*.
Shein, N. 2012. "Newton's Anti-Cartesian Considerations Regarding Space," *History of Philosophy Quarterly* 29(1): 21–38.
Slowik, E. 2009. "Newton's Metaphysics of Space: A 'Tertium Quid' Betwixt Substantivalism and Relationalism, Or Merely a 'God of the (Rational Mechanical) Gaps'?" *Perspectives on Science* 17(4): 429–456.
Slowik, E. 2013. "Newton's Neo-Platonic Ontology of Space," *Foundations of Science* 18: 419–448.
Slowik, E. 2016. *The Deep Metaphysics of Space*. Cham: Springer.
Smeenk, C. and E. Schliesser. 2013. "Newton's *Principia*," in J.Z. Buchwald and R. Fox (eds.) *The Oxford Handbook of the History of Physics*. New York: Oxford University Press. 109–165.
Smolinksi, R. 1999. "The Logic of Millennial Thought: Sir Isaac Newton Among his Contemporaries," in J.E. Force and R.H. Popkin (eds.) *Newton and Religion: Context, Nature, and Influence*. Dordrecht: Kluwer. 259–289.
Snobelen, S.D. 2001. "'God of gods and Lord of lords': The Theology of Isaac Newton's General Scholium to the *Principia*," *Osiris* 16: 169–208.
Snobelen, S.D. 2004. "To Discourse of God: Isaac Newton's Heterodox Theology and His Natural Philosophy," in P. Wood (ed.) *Science and Dissent in England, 1688–1945*. Aldershot: Ashgate. 39–66.
Snobelen, S.D. 2005. "Newton, Socinianism, and 'the One Supreme God'," in M. Mulsow and J. Rohls (eds.) *Socianianism and Arminianism: Anti-Trinitarians, Calvinists, and the Cultural Exchange in Seventeenth-Century Europe*. Leiden: Brill. 241–293.
Snobelen, S.D. 2008. "'Not in the Language of Astronomers': Isaac Newton, the Scriptures, and the Hermeneutics of Accommodation," in S. Mandelbrote and J. van der Meer (eds.) *Nature and Scripture in the Abrahamic Religions: Up to 1700*. Leiden: Brill. 501–540.
Snobelen, S.D. 2021. "Theology in Newton's Study of Alchemy, Chronology and Nature," *Early Science and Medicine* 26(2): 162–180.
Snow, A.J. 1926. *Matter & Gravity in Newton's Physical Philosophy*. New York: Oxford University Press.
Sober, E. 2023. "A Bayesian Double Negative: A Critique of Hume's Treatment of the Design Argument in the *Dialogues Concerning Natural Religion* and A Critique of the Design Argument Itself," in K. Williford (ed.) *Hume's Dialogues Concerning Natural Religion: A Philosophical Appraisal*. New York: Routledge. 47–70.
Soto, C. 2023. "Reconstruyendo la concepción newtoniana de las leyes de la naturaleza," *Revista de las Humanidades de Valparaíso* 23: 309–333.
Sowaal, A. 2004. "Cartesian Bodies," *Canadian Journal of Philosophy* 34(2): 217–240.
Sowaal, A. 2005. "Idealism and Cartesian Motion," in A.J. Nelson (ed.) *A Companion to Rationalism*. Malden, MA: Blackwell. 250–261.
Spargo, P.E. 2005. "Investigating the Site of Newton's Laboratory in Trinity College, Cambridge," *South African Journal of Science* 101(7): 315–321.
Sprat, T. 1667. *The History of the Royal-Society of London*. London.

Stahl, D. 1645. *Axiomata Philosophica.* Cambridge.

Stanley, T. 1655. *The History of Philosophy, Volume I.* London.

Stein, H. 1970. "On the Notion of a Field in Newton, Maxwell, and Beyond," in R.H. Steuwer (ed.) *Historical and Philosophical Perspectives of Science.* New York: Gordon and Breach. 264–287.

Stein, H. 1990. "On Locke, 'the Great Huygenius, and the incomparable Mr. Newton,'" in P. Bricker and R.I.G. Hughes (eds.) *Philosophical Perspectives on Newtonian Science.* Cambridge, MA: MIT Press. 17–47.

Stein, H. 2002. "Newton's Metaphysics," in I.B. Cohen and G.E. Smith (eds.) *The Cambridge Companion to Newton.* Cambridge: Cambridge University Press. 256–307.

Stein, H. 2023. "On Metaphysics and Method in Newton," in M. Stan and C. Smeenk (eds.) *Theory, Evidence, Data: Themes from George E. Smith.* Cham: Springer. 115–138.

Stewart, I.G. 2001. "'Fleshy Books': Isaac Barrow and the Oratorical Critique of Cartesian Natural Philosophy," *History of Universities* 16(1): 35–102.

Tamny, M. 1976. "The Early Epistemological Thought of Isaac Newton," PhD Dissertation. City University of New York.

Tamny, M. 1979. "Newton, Creation, and Perception," *Isis* 70(1): 48–58.

Thackray, A. 1970. *Atoms and Powers: An Essay on Newtonian Matter-Theory and the Development of Chemistry.* Cambridge, MA: Harvard University Press.

Thomas, E. 2018. *Absolute Time: Rifts in Early Modern British Metaphysics.* New York: Oxford University Press.

Thomas, E. 2020. "Descartes on the Animal Within, and the Animals Without," *Canadian Journal of Philosophy* 50(8): 999–1014.

Tuchańska, B. 1988. "Newton's Discovery of Gravity," in G.V. Coyne S.J., H. Heller, and J. Życiński (eds.) *Newton and the New Direction in Science.* Vatican City: Specola Vaticana. 45–53.

Turner, J. 1685. *A Discourse Concerning the Messias.* London.

Tuttle, J. 2022. "Durand and Suárez on Divine Causation," in G.E. Ganssle (ed.) *Philosophical Essays on Divine Causation.* New York: Routledge. 82–101.

Vida, G. 2019. "From *sensorium hominis* to *sensorium Dei*," *Journal of Early Modern Studies* 8(2): 139–149.

Wallace, W.A. 1988. "Newton's Early Writings: Beginnings of a New Direction," in G.V. Coyne S.J., H. Heller, and J. Życiński (eds.) *Newton and the New Direction in Science.* Vatican City: Specola Vaticana. 23–44.

Wallace, W. 2003. "The Vibrating Nerve Impulse in Newton, Willis and Gassendi: First Steps in a Mechanical Theory of Communication," *Brain and Cognition* 51(1): 66–94.

Walsh, K. 2017. "Principles in Newton's Natural Philosophy," in P. Anstey (ed.) *The Idea of Principles in Early Modern Philosophy: Interdisciplinary Perspectives.* New York: Routledge. 194–223.

Watts, I. 1733. *Philosophical Essays on Various Subjects.* London.

Weemes, J. 1632. *Exercitations Divine.* London.

Weiss, R. 2010. "Creation as Parable in Maimonides' *Guide of the Perplexed*," *Interpretations: A Journal of Political Philosophy* 37(3): 259–279.

Westfall, R.S. 1971. *Force in Newton's Physics.* London: Macdonald & Co.

Westfall, R.S. 1976. "The Changing World of the Newtonian Industry," *Journal of the History of Ideas* 37(1): 175–184.

Westfall, R.S. 1980. *Never at Rest: A Biography of Isaac Newton.* New York: Cambridge University Press.

Westfall, R.S. 1982. "Newton's Theological Manuscripts," in Z. Bechler (ed.) *Contemporary Newtonian Research.* Dordrecht: D. Reidel. 129–143.

Westfall, R.S. 1986. "The Rise of Science and the Decline of Orthodoxy: A Study of Kepler, Descartes, and Newton," in D.C. Lindberg and R.L. Numbers (eds.) *God & Nature: Historical Essays on the Encounter Between Christianity and Science.* Berkeley: University of California Press. 218–237.

Wilde, C.B. 1982. "Matter and Spirit as Natural Symbols in Eighteenth-Century British Natural Philosophy," *British Journal for the History of Science* 15(2): 99–131.

Wilkins, J. 1648. *Mathematicall Magick*. London.
Wilson, C. 2008. *Epicureanism at the Origins of Modernity*. Oxford: Oxford University Press.
Wilson, C. 2019a. "What (else) was behind the Newtonian Rejection of Hypotheses?" in P. Anstey and A. Vanzo (eds.) *Experiment, Speculation, and Religion in Early Modern Philosophy*. New York: Routledge. 158–183.
Wilson, C. 2019b. "The Reception of Newton's Theory of Matter and his Atomism," in H. Pulte and S. Mandelbrote (eds.) *The Reception of Isaac Newton in Europe*. London: Bloomsbury. 433–449.
Wojcik, J.W. 2000. "Pursuing Knowledge: Robert Boyle and Isaac Newton," in M.J. Osler (ed.) *Rethinking the Scientific Revolution*. New York: Cambridge University Press. 183–200.
Wolf, L. 2019. "Clarke's Rejection of Superadded Gravity in the Clarke-Collins Correspondence," *History of Philosophy Quarterly* 36(3): 237–355.
Woolhouse, R.S. 1982. "Reid and Stewart on Lockean Creation," *Journal of the History of Philosophy* 20(1): 84–90.
Worrall, J. 2000. "The Scope, Limits, and Distinctiveness of the Method of 'Deduction from the Phenomena': Some Lessons from Newton's 'Demonstrations' in Optics," *British Journal for the History of Science* 51(1): 45–80.
Yenter, T. 2014. "Clarke Against Spinoza on the Manifest Diversity of the World," *British Journal for the History of Philosophy* 22(2): 260–280.

Index

For the benefit of digital users, indexed terms that span two pages (e.g., 52–53) may, on occasion, appear on only one of those pages.

The manufacturer's authorised representative in the EU for product safety is Oxford University Press España S.A. of el Parque Empresarial San Fernando de Henares, Avenida de Castilla, 2 – 28830 Madrid (www.oup.es/en or product.safety@oup.com). OUP España S.A. also acts as importer into Spain of products made by the manufacturer.

www.ingramcontent.com/pod-product-compliance
Lightning Source LLC
Chambersburg PA
CBHW071933120726
47970CB00064B/245/J
* 9 7 8 0 1 9 2 8 6 7 1 1 7 *